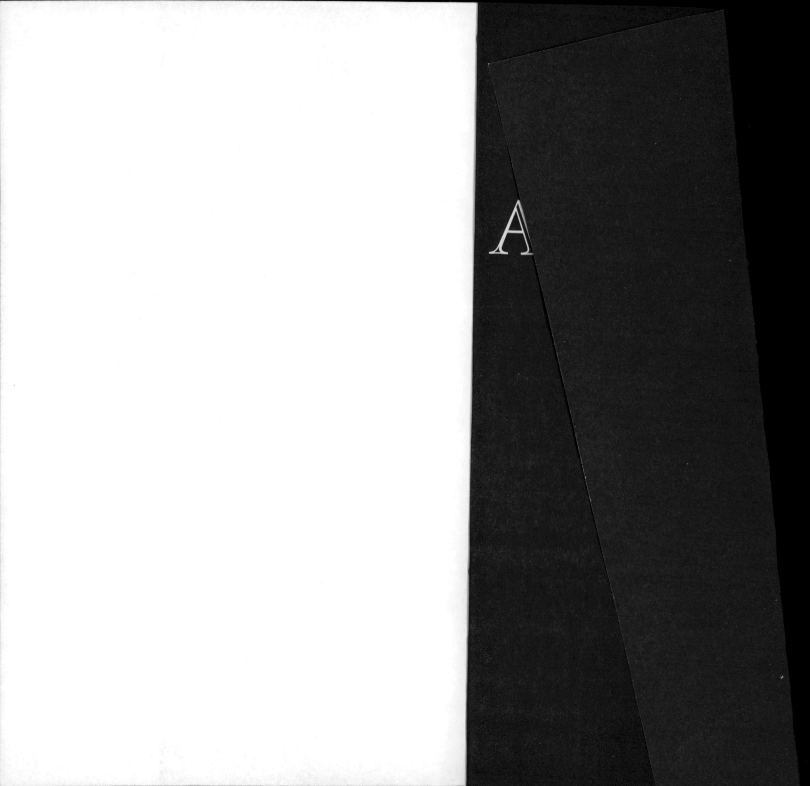

A

ANCIENT SKIES

Constellation Mythology of the Greeks

DAVID WESTON MARSHALL

THE COUNTRYMAN PRESS
A division of W. W. Norton & Company
Independent Publishers Since 1923

For information about permission to reproduce selections from this book,
write to Permissions, The Countryman Press, 500 Fifth Avenue, New York, NY 10110

For information about special discounts for bulk purchases, please contact
W. W. Norton Special Sales at specialsales@wwnorton.com or 800-233-4830

Manufacturing by LSC Communications, Harrisonburg
Book design by Anna Reich
Production manager: Devon Zahn

The Countryman Press
www.countrymanpress.com

A division of W. W. Norton & Company, Inc.
500 Fifth Avenue, New York, NY 10110
www.wwnorton.com

978-1-68268-211-1

10 9 8 7 6 5 4 3 2 1

for Vicki

CONTENTS

INTRODUCTION

STEP OUTSIDE, AWAY FROM THE LIGHT, into the dark and mysterious night. Stand beneath the boundless skies with eyes toward the shining stars. Move your mind beyond the Earth—beyond the here-and-now.

Suddenly, you see a universe of immeasurable size and immense age. You perceive our planet as a cosmic speck orbiting an ordinary star. This star, called the Sun, is only one of billions of stars in our galaxy. And our galaxy is only one of billions of galaxies in the universe.

In the vastness of space, you see some stars so far away that their rays have raced at the speed of light for two million years to reach you. These rays began their journey toward Earth at the time of the birth of humankind; and to glimpse them now is to peer that far into the past. They have traveled farther than the mind can fathom. Yet they, and all the stars that human eyes can see in the sky, occupy only a tiny corner of the cosmos.

As you gaze upon the grandeur, you share in a ritual repeated since the dawn of time. Our ancestors of ages past stood and pondered the same shimmering stars with awe and wonder. One, who lived two thousand years ago, spoke for people of all times when he said that the heavenly light "draws the gaze of mortals upwards, as they marvel at the strange glow through night's darkness, and search, with mind of man, the cause of the divine."[1]

The ancient Greeks stood foremost among those who searched the stars and contemplated the cosmos. Most of them believed that, long ago, the gods created groups of stars, called constellations, to portray themselves in heaven. They thought the gods assigned other stars to depict the deeds of heroes who had faithfully fought on behalf of fellow mortals. Still other stars illustrated tales of divine protection for the humble, or punishment for the haughty. The Greeks concluded that the constellations, as a whole, offered a god-given gift of moral inspiration—an ethical guide for humans to follow.

Some Greeks, like the poet Aratus (315–240 BC), declared that humans, not gods, had contrived the constellations. He explained that "the men-that-are-no-more noted and marked how to group [the stars] in figures and call all by a single name. For it had passed [their] skill to know each single star, or name them one by one."[2]

These words of Aratus provide a profound tribute to our unknown ancestors, who made the earliest efforts to study and understand the heavens. They dared to venture an explanation of the universe, and imparted priceless knowledge and lore. But, sadly, their names have disappeared forever from our collective memory.

The ancient Greeks built upon these timeless ponderings. By the eighth century BC, they had conceived a basic theory of the form of the cosmos. This was recorded by Homer in the *Iliad*— the earliest western writing. In the *Iliad*, the god Hephaestus hammered and forged a shield for the warrior Achilles: "On it he fashioned the earth, on it the heavens, on it the sea, and the unwearied sun, and the moon at the full, and on it all the constellations with which heaven is crowned, the Pleiades and the Hyades and mighty Orion, and the Bear... that circles ever in its place, and watches Orion, and alone has no part in the baths of the Ocean."[3]

In this early age, the Greeks already seemed to conceive of an Earth with surface curvature, if not a complete sphere. They knew the phases of the moon, and the names of certain constellations and asterisms.[4] They noted that one constellation, called the Bear, remained ever bright in the night sky—never dipping below the horizon into the sparkling sea. They marked the celestial North Pole with the Bear's eternal rotation. They perceived that other constellations, as well as the planets, the sun, and the moon, all rose and set beyond Oceanus—that vast body of water thought to encircle the world.

Aratus later described how the stars and constellations "hasten from dawn to night throughout all time."[5] As they travel in unison across the sky, day and night, their relative positions with neighboring stars remain the same; and "all are . . . firmly fixed in the heavens to be the ornaments of the passing night."[6] Aratus and other ancient observers found comfort in the predictable, steadfast nature of the stars. Unlike so many uncertainties that made life on Earth precarious, they could count on the stars to always be where they were supposed to be and behave the way they were supposed to behave.

But among these thousands of reassuring beacons were several stars that seemed to follow no rules. Aratus noted, "Of quite a different class are those five other orbs that intermingle with them and wheel wandering on every side of the twelve figures of the Zodiac," pursuing "a shifty course."[7] These five wanderers, or *planetas* (planets), roamed at will in the sky—capricious and free of the constraints of their fixed-star companions.[8]

Of these wanderers, the morning star named Phosphoros shone forth as the brightest light to bid the night goodbye. Homer described how the day begins when this "star of morning goes out to herald light over the face of the earth—the star after which follows saffron-robed Dawn."[9] On some days, Phosphoros foretold momentous events. Homer praised the morning star for announcing the day that the long-lost Odysseus returned at last to his island kingdom, after an absence of twenty years: "Now when that brightest of stars rose which beyond others comes to herald the light of early Dawn, then it was that the seafaring ship drew near to the island" of Ithaca.[10]

Hesperos, the evening star, rivaled the beauty of Phosphoros. According to Homer, "the star of evening" was "set in heaven as the fairest of all."[11] In fact, the two stars appeared in essence the same, and seemed to take turns lighting the night sky. At certain times of the year, Phosphoros adorned the morning. At other times, Hesperos graced the evening. But the two never revealed themselves on the same nights.

The explanation for this strange behavior had to await the insight of Pythagoras (c. 572–490 BC). The famed astronomer and mathematician discovered that the two were actually one heavenly body that sometimes shone in the morning and other times in the evening.[12] The Greeks renamed the resplendent wandering star for Aphrodite—the goddess of love and beauty.

Later, the Romans called the planet Venus—the second brightest celestial light in the night sky, after the moon. On a clear and moonless night, her glow is capable of casting shadows on Earth. Little wonder that she so amazed ancient observers!

It seemed right that the five radiant, wandering stars—or planets—should be named for the willful gods. Following Aphrodite (Venus) in magnitude came Zeus (Jupiter), the god of sky and storm; Ares (Mars), the blood-red god of war; Cronos (Saturn), the Titan father of Zeus; and Hermes (Mercury), the messenger god that flew fastest in transit across the sky.[13]

These five praiseworthy planets intrigued observers, but they proved too capricious in motion for practical uses. They could not be counted upon for marking cardinal directions, or noting the passage of time and seasons.

Instead, the Greeks depended on the ever-faithful fixed stars— especially those that circumscribed the celestial North Pole, or wheeled within the zodiac. The zodiac, with its twelve constellations, circles the Earth and envelops the ecliptic path of the sun, moon, and planets across the sky. In ancient times, as the sun slowly made its way from one zodiacal constellation to the next, it marked the start of another month, and the passage of time toward another season.

The Greeks defined nine of these twelve constellations as animals. In fact, the name *zodiac* comes from the ancient Greek words *zodion kuklos*, meaning *circle of little animals*.[14] The Greeks also described most of the constellations in the sky as fauna. According to Plato (c. 428–348 BC), "The fixed stars were created, to be divine and eternal animals, ever-abiding and revolving after the same manner and on the same spot . . . circling as in dance."[15]

The forty-eight classical constellations, which comprise the core of our modern star groupings, emerged in near-final form by the fourth century BC. Of these, twenty-five depict animals, including thirteen mammals, three birds, three reptiles, two fishes, one crustacean, two arachnids, and one half-mammal, half-fish (Capricornus). In addition, two constellations portray the mythical half-man, half-horse centaurs. Greek lore honors most of these celestial animals for their earthly service to gods and men.

Humans, on the other hand, figure in only a few of the forty-eight constellations. Moreover, mythology often reveals their marginal worthiness. In contrast, one divine being—Virgo—appears in anthropomorphic form, and in the stories of the stars she is held in highest esteem. So too are several constellations that represent the demigods—the heroic sons of divine fathers and mortal mothers. In addition to these, nine inanimate objects make up the remaining constellations.

While ancient observers defined most of the constellations as animals, they also portrayed them in the most important places along the celestial sphere. Furthermore, they defined animals in greater starry detail.[16] Human and anthropomorphic anatomy received less representation.[17]

For example, prominent stars marked animal facial features.

But they remained notably missing from most human faces. These faceless humans seem strangely reminiscent of the simplified human stick-figures fashioned in charcoal on walls of Ice Age caves. Alongside these stick-figures, magnificent bison and beasts stampede in splendid detail and vibrant colors of ochre.

Both examples—the constellations and cave paintings—suggest an early human reverence for animals. This, perhaps, was due to their greater prowess, or their importance as a source of provision. Only later did human dominance and domestication of animals lead to belief in their brutish inferiority. Nevertheless, throughout the ages, animals retain their prominent place in the sky.

Many of these animals, and certain humans and gods, had long been the subject of prehistoric stories and oral traditions. Now they gained greater fame as generations of storytellers envisioned them in the stars—assigning them tangible forms as constellations.[18]

Through the ages, the night sky came to life with animation and drama. But the stories in the stars provided more than simple entertainment. They offered a first essential step toward scientific astronomy by prompting speculation on the matter and meaning of the mysterious heavens.[19]

After all, science begins with awe and wonder of the unknown. Wonder leads to speculation. Speculation generates theory. Theory is subject to critical analysis and further investigation in a relentless search for facts. This might involve thousands of single contributions during thousands of years. Mythology has often served as a catalyst for this scientific process.

Speculation on the nature of the universe gave birth not just to science, but to philosophy. In ancient Greece, astronomers who contemplated the heavens in the broadest terms became the first philosophers.[20] They concluded that the star-studded cosmos, with its predictable and rational motions around the Earth, represented the farthest reaches of an orderly, harmonious whole—a universal oneness.

Thales of Miletus, the famed astronomer and first western philosopher, embraced this holistic approach. Soon he searched for a single substance that surely served as the primal basis of the universe. He concluded that water was the elemental source, while his younger contemporary, Anaximenes, made the same case for air.

Still others, including Anaximander, Pherecydes, Xenophanes, and Heraclitus, perceived of an abstract origin. They envisioned an infinite, indefinable reality—a single cause and source from which all things emerge. Their successors, the three Athenian philosophers—Socrates, Plato, and Aristotle—also embraced this belief. They concluded that this divine reality is tangible in the natural world just as it is intangible in the human mind and soul— as one harmonious whole.[21]

Greeks believed that to contemplate the natural world, to seek to understand it, and to glorify it in the arts were worthy endeavors that directed the mind toward a higher reality. In accordance with this belief, classical works of art centered on a simple and sublime object that drew attention beyond itself to a larger context—to an ideal greater than itself. Sculptors, for example, sought to fashion figures that achieved realism, then moved beyond reality to a transcendent grace and beauty. Their art represented the real reaching toward the ideal, or nature reaching toward spiritual perfection.

This concept of the real reaching toward the ideal became the standard for individual achievement as well. Greeks used the word *arete*—personal excellence and virtue—to describe the appropriate pursuit of life to its full physical, mental, and spiritual potential. Several constellations commemorated those who achieved this goal.

In the physical sense, *arete* alluded to athleticism, strength, and beauty. In the mental sense, it involved the attainment of knowledge and wisdom. In the spiritual sense, it meant goodness, fairness, and humble devotion. In the age of Socrates, Plato, and Aristotle, *arete* also involved the attainment of spiritual truth through contemplation.[22]

Ideally, a person would seek a balance of physical, mental, and spiritual merits, rather than excelling in one at the expense of the others. An obsession with the physical might create a narcissist; the mental a pedant; the spiritual a zealot. In order to possess the proper proportion, a person was wise to follow the two-part advice codified by the earliest western philosophers.

This sage advice was inscribed on a wall of Apollo's temple at Delphi. The first part said: "Know thyself!" Or, be introspective, and know your strengths and weaknesses in order to work toward improvement. The second said: "Nothing in excess!" Or, maintain harmonious balance in all things.[23]

This same search for *arete*, in proper proportion, also found application on the community level. Religious festivals featured contests in athletics, music, drama, and the visual arts. The Greeks would not have thought to splinter these into separate occasions, because they contributed collectively to a higher ideal.

Tragic drama, for example, remained ever aligned with a well-established code of ethics. Performances typically portrayed the theme of divine punishment for excessive pride. The same theme prevailed in the tales of the constellations. In both cases, the stories evinced a longstanding belief in the moral order of the universe.

In the prehistoric past, Greek spiritual awareness was animistic. The Greeks believed that all natural features and phenomena possessed an indwelling spirit. In time, animism gave rise to polytheism, which assigned anthropomorphic forms and personalities to many of these same features and forces of nature. Thus, while animists believed that thunderstorms embodied a powerful spirit, polytheists defined the spirit as Zeus—a god with manlike traits.

But Zeus, and the other gods and spirits, still remained subject to an infinite, indefinable creative force—a divine oneness, a harmonious whole—from which all things arose. This belief in a single creative force is a notable feature of ancient cultures around the globe.[24] In Greece, the prehistoric belief emerged in the earliest written records of philosophy and mythology. The works of Hesiod, Anaximander, Plato, and others attest to this.

These thinkers flourished in an environment of free inquiry that allowed them to discuss and disseminate their beliefs. They experienced the advantage of living beyond the shadow of deep-seated despotism. The entrenched empires of the Near East, backed by aristocratic priesthoods, had used the weapon of fear since the dawn of civilization to keep the masses simple and submissive.

But the Greeks had little patience for despots and priests. Instead, they dabbled in democracy, and practiced a personal religion that allowed them to escape the onus of a powerful priesthood. Priests remained in the role of temple officials, while real worship occurred at altars and holy sites in homes, fields, forests, and public places. Each person could communicate directly with the divine through prayers and dreams, or with offerings of food, water, wine, and incense. Festivals, processionals, and performances offered further features of their informal religion.

The early Greeks acceded to no sacred law or scripture—no creed or doctrine. A menagerie of myths—variously interpreted as profound or petty—stood in the place of formal theology, and lack of dogma meant freedom of conscience.

In this environment of free inquiry, philosophers and scientists discovered new paths of learning. They searched for rational explanations of natural phenomena, even while proclaiming the spiritual essence in all things. They never came to the vain conclusion that a thing did not exist simply because they could not probe or comprehend it. For them, the world and the cosmos became compelling rather than menacing. It became a place of beauty rather than fear.

The Greeks certainly differed in their provocative opinions. They often quarreled and even warred among themselves. But they also showed a striking degree of toleration. They glorified singular achievements in philosophy, science, the arts, and other areas that directed attention toward a higher ideal.

The very atmosphere inspired intellectual vitality, curiosity, and elevation of the mind and soul. Socrates provided a striking example of the impulse of the time. As a seeker of truth, he asked penetrating questions rather than dictating his beliefs. He readily rejected conventional thought in his tireless search for enlightenment.

This incessant soul-searching, and dismissal of social assumptions, caused many Greeks to redefine their opinions of the gods. In the past, they had praised the deities for ruling fairly from their pantheon on Mount Olympus. But by the fifth century BC, the gods had lost much of their charm. Popular myths demoted them to a soap opera of family squabbles, romantic trysts, and shocking intrigues that portrayed them as prone to petty jealousies rather than divine justice.[25]

As the status of deities declined, that of humans ascended. Soon the two resembled each other in villainy and nobility. The gods seemed more like illustrious ancestors than supreme beings. For they, too, stood subject to a universal truth and moral code that so often exposed their folly. By the Classical Period (510–323 BC), mythology tended to emphasize the humanlike frailties of gods, while lauding the godlike achievements of humans.

This mythology was well represented in the celestial panorama. The forty-eight classical constellations, and their associated stories, depicted human and divine interaction that portrayed a code of ethics—a lesson for proper living. The code may be paraphrased as follows: Honor and serve gods and fellow mortals to promote peace and harmony, and avoid arrogance and greed that breed divisiveness and strife.

The constellations consistently conveyed this code by depicting love and devotion among deities and mortals. In contrast, several constellations stressed the negative effects of violating the moral code, and recounted the catastrophic results of vanity and avarice.

The ancient skies came to life with tales of love and devotion and of arrogance and greed. They portrayed spirited passions, bold adventures, exuberant joy, and tragic sorrow, all of which rose and fell on a rolling wave of emotions. These stories voiced the same confident expression that animated tragic and comedic drama, lyric and epic poetry, art, architecture, science, and philosophy.

In addition to impassioned stories and moral lessons, the skies provided a practical guide to the cardinal directions and the passage of time.[26] The ancient Greeks believed that the gods gave humans the sun by which to awaken and work. They gave the stars and constellations to tell them when to plant and harvest, when to sail, and when to avoid the stormy seas.[27]

The movement of the sun and its shadows offered a daytime clock. The stars likewise marked the passage of night. Homer's *Odyssey* twice noted the "third watch of the night," when "the stars had turned their course" around the celestial North Pole. From this, weary watchmen knew that dawn was only a few hours away.[28]

Five centuries later, Aratus described how to use the "twelve signs of the zodiac . . . to tell the limits of the night."[29] By watching these constellations as they slowly rose over the eastern horizon, a person could determine the time of approaching daybreak. For, as Aratus declared, "the sun himself rises," always with one of the constellations. [30]

Classical Greek star lore remained forever enlightening and practical. It was not astrological. Astrology was a Near Eastern invention that only later arrived in the West, after Greek expansion into Asia. [31]

The Classical era ended with this invasion, as Alexander the Great launched his conquests in the fourth century BC. In the two millennia that followed, empires rose and fell, a hundred generations lived and died, and a heartbreaking amount of ancient learning was lost. But the tales of the forty-eight constellations still remain—in memory of the glory of ancient Greece.

Atlas holding a celestial globe. Sculpture in marble (c. 150 AD) copied from a Hellenistic original of the second century BC. Farnese Collection, National Archeological Museum, Naples, Italy. The Farnese Atlas globe provides the earliest extant images of the Greek constellations. Celestial globe images always appear in reverse, as if viewed from outside the celestial sphere.[32]

PART ONE

—

THE ANCIENT
CONSTELLATIONS

The following six chapters comprise the stories of the forty-eight classical constellations. These stories emerged from the works of numerous authors who lived in Hellas (Greece) during a period of nine centuries—from Homer (c. 750 BC) to Claudius Ptolemy (c. 150 AD).

The chapters reflect an ancient Hellenic perspective and provide literal translations of Hellenic names for constellations and principal stars. The names are capitalized in the text and appear in all-capital letters when first referring to the constellation. The constellation stories fall conveniently into several subject areas that conveyed a code of ethics to ancient audiences. The six chapters reflect these themes.

The adjoining hemisphere charts illustrate the ancient Hellenic manner of viewing the celestial sphere. The southern hemisphere chart shows an empty expanse around the celestial South

Pole. This is because the southern sky was too far south for observations from Mediterranean latitudes.

Each individual constellation chart portrays a starry image reconstructed from the primary stars defined by ancient observers. Each chart also provides the original constellation name in Greek, the literal English translation, and the modern name in Latin with pronunciation. In addition, the name of the primary star appears, with its celestial coordinates of right ascension and declination, based on epoch J2000. These coordinates assist the reader to find the constellation in the sky. Finally, the chart presents the constellation's corresponding modern star designations and names.

Also appearing throughout are twenty-eight plates from Alexander Jamieson's *Celestial Atlas* (1822). The Jamieson images are in the public domain and available from the United States Naval Observatory in Washington, DC, and on the USNO Library website.

Capricornus depicted in the *Uranometria* of Johann Bayer, 1603. Courtesy of the Museum of History and Science Library, Oxford, England. Photo by author.

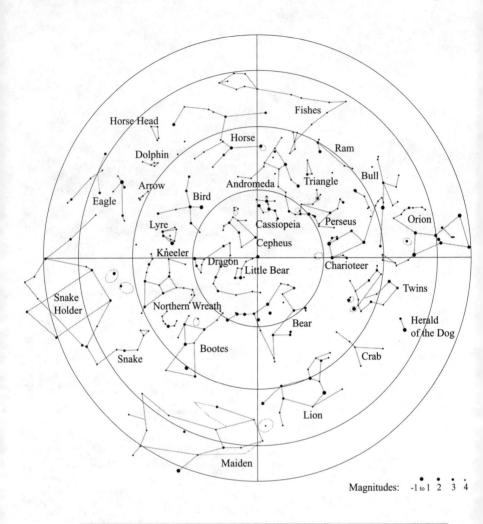

Horse Head
Dolphin
Arrow
Eagle
Lyre
Kneeler
Snake Holder
Snake
Bootes
Northern Wreath
Dragon
Little Bear
Cepheus
Cassiopeia
Andromeda
Horse
Bird
Triangle
Fishes
Ram
Bull
Perseus
Orion
Charioteer
Twins
Herald of the Dog
Bear
Crab
Lion
Maiden

Magnitudes: -1 to 1 2 3 4

Ancient Northern Hemisphere

Sea Monster

Water Bearer

Southern
Fish

Goat Horn

River

Southern
Wreath

Archer

Incense
Altar

Hare

Argo

Wild Animal

Scorpion

Dog

Centaur

Claws

Hydra

Crow

Crater

Magnitudes: -1 to 1   2   3   4

Ancient Southern Hemisphere

1

THE VALUE OF DEVOTION

CONSTELLATIONS
—
CHARIOTEER (Auriga)
GOAT HORN (Capricornus)
FISHES (Pisces)
SOUTHERN FISH (Piscis Austrinus)
DOLPHIN (Delphinus)
EAGLE (Aquila)
WATER BEARER (Aquarius)
INCENSE ALTAR (Ara)

STARS
GOAT (Capella)
KIDS (Haedus 1, Haedus 2)
DONKEYS (Asellus Borealis, Asellus Australis)

Left: Jamieson Plate 4: Auriga

From dusk to dawn, thousands of shining stars rise in the east and roam westward through the night. When you search the skies, with eyes alert to the tiny beams of light, you see a starry parade, in steady procession, traveling in family groups called constellations.

As you ponder the luminous heaven, listen to the tales of ancient skies—the tales of old, of heroes and foes—conveyed by the constellations. They tell us to move in harmony, like the stars above. They teach us to walk in peace, as one, and avoid conceit, greed, and strife. They reveal the path to a complete and contented life.

ERECHTHEUS AND ATHENA

Long ago, an Athenian named Erechtheus gained great fame by seeking a life that was whole in body, mind, and soul. In his youth, all of Athens adored him as an amazing athlete. He ran swifter of foot than any man, and on horseback had no rivals. Still, Erechtheus saw room for improvement and determined to search for greater speed.

First, he carefully matched his four fastest steeds. Then he trained them as a team to pull a two-wheeled cart of his own design. Together they rushed down dusty roads—faster than man could imagine. Farmers with ox carts—slow as snails—gaped as the charioteer flew past with his hair and tunic trailing.

All the while, Erechtheus perfected the chariot's wheels and learned to manage the team on tight turns and rolling hills. Others followed his lead and raced by his side, like the rush of the wind. But none ever matched his valor or speed in the Panathenaic or Olympian Games.

Far beyond being an athlete and clever inventor, Erechtheus became the most pious of men—a devoted disciple of the goddess

Athena. He was the first to take a torch in hand and lead a procession, under the stars, to the limestone crest of the Acropolis. There he erected the first Athenian temple built in her honor.

Athena admired Erechtheus with fond affection, and he responded to her in kind. In this way, he reached the height of human potential. For we can only gain greatness through love and humble devotion toward an ideal greater than ourselves. As an accomplished athlete, and a wise and pious person, he achieved *arete*—excellence and virtue; and he proved that *arete* never results in arrogance, but only in adoration toward others.

At the end of his long and illustrious life, Erechtheus attained a reward in heaven. Here he shines in the night to show what mortals may become.[1] As a cluster of stars, he appears as the CHARIOTEER, standing tall and tugging the reins with the same powerful poise as the marble statue in Delphi.[2]

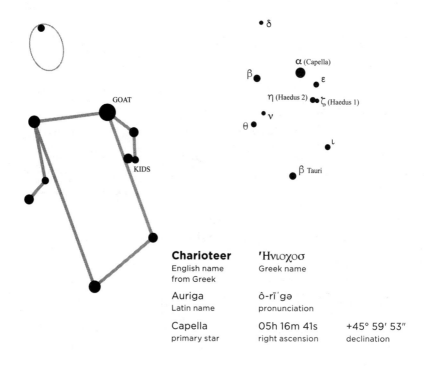

Charioteer	'Ηνιοχοσ	
English name from Greek	Greek name	
Auriga	ô-rī'gə	
Latin name	pronunciation	
Capella	05h 16m 41s	+45° 59' 53"
primary star	right ascension	declination

Charioteer at Delphi. Sculpture in bronze, c. 470 BC. Courtesy of the Archeological Museum, Delphi, Greece. Photo by author. The intense, erect posture matches that of the constellation figure.

Through the centuries, his children's children followed his example, and built more temples to revere their beloved Athena. Sometimes they named them after Erechtheus, to commemorate his devotion. The most recent—the Erechtheum—rises on the Acropolis near the Parthenon, the temple of the maiden Athena.

The Erechtheum shelters the goddess's sacred tree—the olive— the very one she caused to sprout on the Acropolis long ago, when first she pledged to protect the city. The temple also houses an olive bough, carefully carved in Athena's immortal image.

As the craftsmen constructed the Erechtheum, they fashioned a set of six columns in the form of marble maidens to oversee the sacred tree. These maidens—the Caryatids—are priestesses of the goddess Carya—the guardian of walnut, hazelnut, and related fruit-bearing trees.[3] It was Carya who furnished these fruits to save our ancestors from hunger when times were hard, long ago, before humans knew how to herd or farm. Now her maidens protect the olive—Athena's greatest gift to mortals.

In life, Erechtheus adored Athena for all her gifts and blessings. But he also loved all beings, great and small. Because of this, as a constellation, he shows gentle kindness by bearing on his left shoulder an old and feeble nanny GOAT named Amalthea. The Goat shines as one of the brightest stars in the sky because she, too, was known for her service to mortals and gods—including to Zeus.

ZEUS AND COMPANIONS

When Zeus was newly born, his mother, Rhea, laid the tiny babe in a lonely cave on the isle of Crete. Here she hoped to hide him from his fearsome father. Cronos—the Titan father—planned to destroy Zeus and the other infant gods he had sired. In this way, he sought to forestall the prophecy that one of them would one day depose him. Rhea, desperate to save her son, relinquished Zeus to the care of Amalthea, to nurse and raise him in seclusion.

The nanny Goat dutifully adopted Zeus—not from fear, but from affection for the goddess Rhea. And so Amalthea suckled the infant god alongside her own twin Kids—her rambunctious baby goats. Soon she came to love him as her own. Zeus spent his childhood in Amalthea's cozy cave, and grew to youth without the soft touch of his mother. But Amalthea cuddled and cared for him dearly, and delighted to see him befriend the animals and nymphs all around him.

Kindhearted Amalthea also raised an orphan goat beside her own twin Kids and the infant Zeus. She named him Aegoceros—meaning Goat Horn—because he loved to lock horns with the other goats, and push and play. The four grew up as brothers and remained fast friends for life. One can only imagine the adventures and escapades of the three young goats and the toddler god as they explored every deep, dark nook and cranny of their cavernous home.

As they grew older, they ventured into the outside world to scramble over boulders and splash across sparkling brooks. From morning to night, they played their favorite games, or dared each other to perform reckless deeds, as boys will do. No doubt their poor nanny worried whenever they wandered from sight.

As the little foursome roamed the rocky shore one blustery day, Aegoceros—an excellent swimmer—dove into the spray and swam deep into the aquamarine sea. After a long absence that caused his companions to anxiously watch and wonder, he suddenly bobbed to the surface. High above his head he held an exquisite conch shell that glistened like gold in the morning light.

No sooner had he splashed ashore to the pebble beach than he placed the seashell against his lips and produced a perfectly deafening trumpet blast. His wide-eyed partners jumped in alarm, then leapt and laughed in delight. Falling in line, they formed a parade, marching and shouting, and taking turns bugling until blue in the face. From that day forward, Aegoceros was never seen without his shell dangling by a string at his side.

The four friends enjoyed a carefree childhood, but all boys must grow up to assume the duties of adults. The time at last arrived for Zeus to claim his divine destiny—to rule over gods and Titans as the prophecy foretold.

Beside him stood his constant companion—Aegoceros—who followed him into battle against Cronos and the Titan host. The gods Hephaestus and Dionysus, riding on two Donkeys, also joined the march; and other deities, satyrs, and sympathetic supporters likewise took up arms as soldiers.

After a long and weary journey, the army of Zeus discovered the dreaded lair of the Titans, who remained unaware of their presence. Silently, Zeus and his band crept forward as close as they dared. In the final moment, near the mouth of the Titan cave, Aegoceros put his lips to the conch shell and gave a mighty blast. The deafening sound reverberated through the winding cavern and echoed off surrounding hills, causing the two Donkeys to bray with all the strength of their lungs. The awful din so unnerved the Titans that they dropped their weapons and fled in panic, never pausing to look behind them. Zeus and his faithful followers celebrated a great victory that night, although many battles remained to be fought.

The hero of the moment was Aegoceros, who Zeus—the god of sky and storm—now honored in grateful embrace. Again and again, in battles to come, the rustic goat would turn the tide in favor of the gods, not by the strength of the sword or by dazzling displays of daring, but with a simple conch shell—a childhood toy—from the depths of the deep blue sea.

While the years passed and Zeus waxed strong as ruler of the gods, Amalthea, his nanny, waned in frail old age. When the day arrived that the nanny died, Zeus wept tears that showered down in heavy rains upon the earth. To remember Amalthea from that moment on, he kept her hide as a soft and supple sign of her love and protection. In desperate times, he held it close and bore it in battle with his Titan foes. The hide became his breastplate—the aegis—his symbol of invincible strength. Also, in grateful esteem, he gave the Goat a star of her own in the sky, where Erechtheus carefully carries her on his shoulder.

Zeus, being immortal, remained in his prime while most of his friends finally dwindled and died. Even his playmates, Amalthea's twin KIDS, followed after their mother. Now they rest in Erechtheus' left hand, close beneath the nanny, where they shine as two tiny, twinkling beams of light.

In the same way, Aegoceros came to the end of his days. High in heaven, his constellation, called GOAT HORN, honors him for bearing the bugle that won so many battles. It also commemorates his aquatic skills and shows him swimming through the stars, like he once splashed in the Aegean Sea. But now he has a fish's tail to propel him along as he holds his head and horns high above the surf.

The two DONKEYS, who aided the gods as beasts of burden and helped to rout the Titans, also share a prominent place above. Here they appear as two neighboring astral lights, riding eternally atop the head of the constellation called the Crab. The Crab, as punishment for bad behavior, now serves the Donkeys as *their* beast of burden. To add to their reward, a heaping

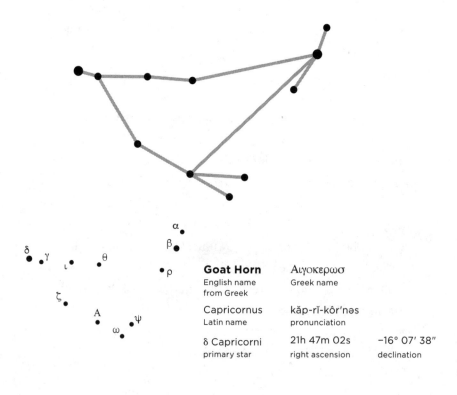

α
β
δ γ
  ι   θ
ρ
ζ
A   ψ
 ω

Goat Horn
English name
from Greek

Αιγοκερωσ
Greek name

Capricornus
Latin name

kăp-rĭ-kôr′nəs
pronunciation

δ Capricorni
primary star

21h 47m 02s
right ascension

−16° 07′ 38″
declination

Jamieson Plate 21: Capricornus, Aquarius

manger of golden straw rests within reach of their muzzles. At
night, the straw can be seen shimmering in the distant sky as
a hazy cluster of stars, with the hungry Donkeys nibbling from
either side.[4]

THE ESCAPE OF APHRODITE AND EROS

Zeus and his friends and followers won the first fights with the
Titans. But the conflict turned against the gods when Gaea, who
is Mother Earth and mother of the Titans, brought forth her final

offspring—the monster Typhon. The creature's enormous size, earth-shattering strength, and fiery breath made him the fiercest beast to ever be unleashed. Typhon soon rumbled forth from the bowels of the earth, determined to hunt down the deities and devour them one by one. Several fled in despair, or disguised themselves as birds or fishes in order to take flight in the sky or find refuge in the sea.

The goddess Aphrodite and her son, Eros, hid in the land of Syria. There they thought themselves safely beyond his reach. But as they dawdled dreamily down the bank of the Euphrates River one day, Typhon appeared without warning, as was his way. With a deafening roar, he burst from the earth and pursued them in volcanic fury. Suddenly pinned between fire and water, Aphrodite grabbed Eros in her arms, and, in desperation, dove for the depths of the river.

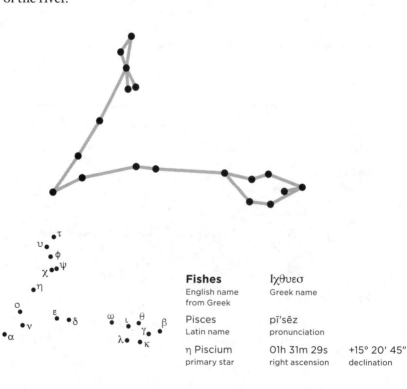

Fishes   Ιχθυεσ
English name   Greek name
from Greek

Pisces   pī'sēz
Latin name   pronunciation

η Piscium   01h 31m 29s   +15° 20' 45"
primary star   right ascension   declination

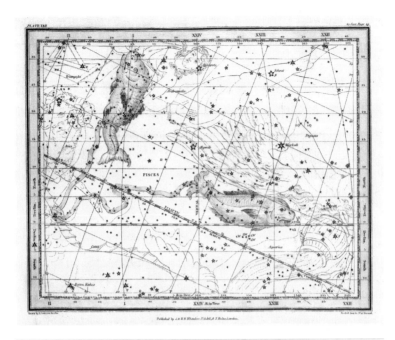

Jamieson Plate 22: Pisces

Aphrodite was born of the sea, and easily adapted herself and her son to the water. In haste, they took the form of fishes in order to breathe and swim in their new surroundings. To keep her son close by her side, she tied each end of a cord to their tails.

Fortunately for them, Typhon was a fiery fiend who despised the cold and quenching effects of river and sea. He balked at the thought of pursuing them as they darted among the slippery rocks, through tangles and snags, and swam with the current to safety. Steaming and hissing in bitter rage, he angrily abandoned the chase. A comfortable distance downstream, Aphrodite and Eros emerged from their watery refuge and dried themselves in the springtime sun and balmy breeze.

In the years that followed, Syrians who lived along the river refused to eat fish, to show their respect for the divine mother and child. Zeus, for his part, celebrated the escape of the two by designating stars to form the FISHES. High in the night sky, Aphrodite is seen as she swims the Euphrates' flowing stream, still tied by a cord to Eros, who dashes and darts to the left.[5]

As a further distinction, Zeus ordained that the sun should traverse their constellation in its annual passage across the sky. Thus, every year, when the sun aligns with the Fishes, it marks the close of the cold and rainy season, and heralds the emergence of life-giving spring.

THE FISH THAT SAVED A GODDESS

In later years, another fish of the swift Euphrates gained similar fame for saving Derceto, a daughter of Aphrodite. Derceto, in honor of her mother's escape from Typhon, had always favored the fishes of the river. At her temple in Bambyce, near the west bank of the Euphrates, she kept her favorite scaly friends in a sacred lake. Each day, she fed them as they gathered in schools and gaped above the surface.

One evening, as she leaned with food in hand far over the water, she lost her footing and slipped into the glassy lake. Quick as a flash, the largest fish flicked his fins and darted beneath her sinking form. Rising to the surface, he laid her on the bank and saved Derceto from drowning.

Aphrodite, the grateful mother, rejoiced in the rescue and recognized the valiant deed by assigning the fish a place in the cosmos. Here, he is known as the SOUTHERN FISH because of his position low in the southern sky. Like Aphrodite and Eros, he, too, is revered by the Syrians, who honor the three by fashioning delicate fish designs out of silver and gold.

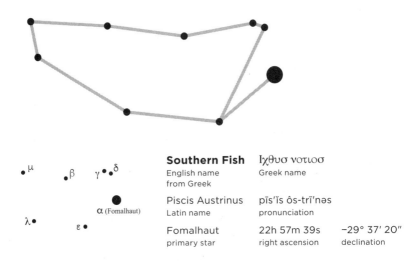

Southern Fish
English name
from Greek

Ιχθυσ νοτιοσ
Greek name

Piscis Austrinus
Latin name

pīs'ĭs ôs-trī'nəs
pronunciation

α (Fomalhaut)

Fomalhaut
primary star

22h 57m 39s
right ascension

−29° 37' 20"
declination

The Southern Fish received further reward by being adorned with one of the most dazzling stars in heaven. It marks his gaping mouth beneath two tiny eyes. As he swims across the celestial sphere, he opens wide to drink a draft of water poured as a pure libation by the constellation called the Water Bearer.[6]

POSEIDON'S LOYAL DOLPHIN

Another wanderer of the watery deep—the Dolphin—received similar celestial acclaim for his selfless service to a divine being. Poseidon, god of the sea, presided over all the dolphins and other creatures of his aquatic domain. But his powerful presence caused his minions to shudder and stay away. As a result, he remained quite lonely as he pined away in his palatial undersea cave.

At last, he declared his longing desire to marry a modest nymph named Amphitrite. The coy sea-maiden was flustered and frightened by the affections of the powerful god. Straightaway she fled far to the west, where the waters of the Mediterranean lap the shores near the Atlas Mountains.

Poseidon confided his feelings to his friend—the Dolphin—and convinced him of his genuine love for the girl. The Dolphin, fully assured of the god's good intentions, departed with haste through the salty sea to search for Amphitrite.

High and low he looked—not in wide open waters, nor on islands where sea folks gather, but in secret coves and sunken caverns far beyond the horizon. He knew the hideouts of the deep—every crack and chasm, every cave and coral reef. Time and again, as he swam beneath the waves and leaped above the surf, he scattered schools of fishes and slipped away from menacing sharks.

After searching for many days, the Dolphin discovered Amphitrite, trembling in a tiny cave, surrounded by a small gathering of seahorses and starfishes, her only friends. With gentle words of encouragement, he assured the nymph of her suitor's love and convinced her to wed the watery god. Upon their long-awaited return,

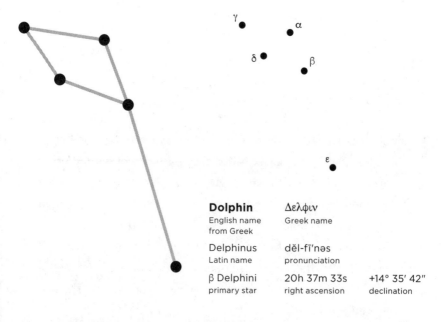

Dolphin
English name
from Greek

Δελφιν
Greek name

Delphinus
Latin name

dĕl-fī'nəs
pronunciation

β Delphini
primary star

20h 37m 33s
right ascension

+14° 35′ 42″
declination

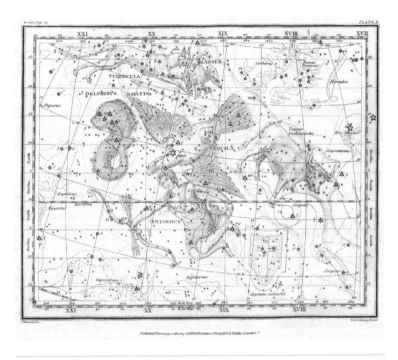

Jamieson Plate 10: Delphinus, Aquila, Sagitta

he presented the timid bride to Poseidon. The Dolphin then presided over the nuptials before a beaming, bubbling audience—a colorful array of the creatures of the sea.

On countless other occasions, the Dolphin and his kinsmen served humans equally well by guiding ships and saving floundering seamen. Even the Argonauts fondly recalled how "dolphins in calm weather leap up from the sea and circle a ship in schools as it speeds along, sometimes showing up in front, sometimes behind, sometimes alongside, and joy comes to the sailors."[7]

For his service to Poseidon, and all of mankind, the DOLPHIN earned an eternal home in the stars. There he leaps with elation and offers a sign of assurance at night for navigators traveling on the trackless sea.

THE MIGHTY EAGLE OF ZEUS

Mortal men, like Erechtheus, and beasts of the earth, like Amalthea, the Kids, Aegoceros, and the Donkeys, faithfully served the divine deities. The Syrians, too, revered Aphrodite and Eros as the Fishes. And watery creatures, like the Southern Fish and the Dolphin, aided and honored the gods.

In the same way, the birds of the air displayed their devotion on many occasions. In fact, all the deities have their favorite birds that faithfully serve them. The little owl is Athena's constant companion, as shown on the silver drachmas of the Athenians. Zeus's winged friend is the mighty Eagle.

As Zeus's campaign continued against the Titans, the war clouds gathered. Lightning flashed in the stormy sky and rumbled in echoes of thunder. Beneath the terrible tempest, Zeus prepared his anxious army for battle. As he attempted to quell their fears and prompted them to perform valiant deeds, an enormous Eagle suddenly appeared. Like a rush of wind, the majestic bird alighted by Zeus's right side.

Athenian drachma, minted from silver of the Laurium mines in the fifth century BC. The obverse shows the head of Athena. The reverse depicts her symbols—the owl and olive sprig. Marshall Collection, Loma Paloma, Texas. Photo by author.

A spirited shout resounded through the ranks as the soldiers witnessed this favorable omen. Their hearts now swelled with courage, and they bolted forth to scatter the Titan foe. As they hammered swords against shields, clinking and clanging in wrathful rage, the Eagle circled high above, offering inspiration.

At the end of the day, with the battle won, the Eagle returned to the god of sky and storm. Zeus raised the raptor high above his head, perched on his upraised fist, and declared him his friend from that day forward.

In ages to come, the Eagle often proved his allegiance and favored Zeus with offerings of esteem. When Zeus demanded a cupbearer—a person of grandeur and grace worthy to bear libations to the gods, the Eagle brought him Ganymede—"the fairest of mortal men."[8] To the heights of Olympus he carried the youth, carefully cradled in his massive claws.

For this thoughtful deed, and to praise all of his kind, Zeus immortalized the EAGLE, allowing him to soar in the nighttime

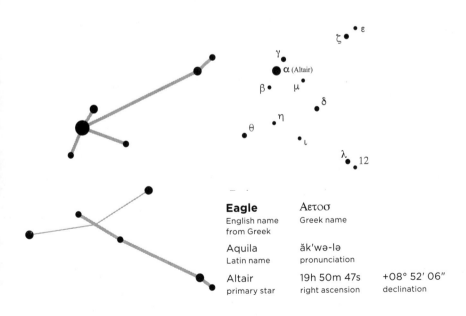

Eagle
English name
from Greek

Αετoσ
Greek name

Aquila
Latin name

ăk'wə-lə
pronunciation

Altair
primary star

19h 50m 47s
right ascension

+08° 52' 06"
declination

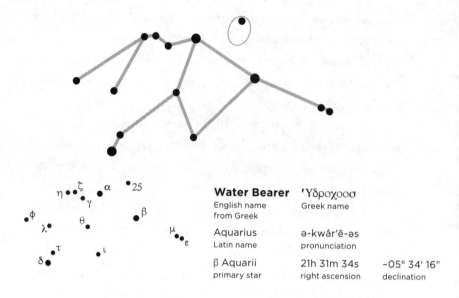

Water Bearer 'Υδροχοοσ
English name | Greek name
from Greek |

Aquarius | ə-kwâr'ē-əs
Latin name | pronunciation

β Aquarii | 21h 31m 34s | -05° 34' 16"
primary star | right ascension | declination

sky—in splendor for all to see. His breast is turned toward the Earth and he bears a bright star on his shoulders. Three neighboring astral lights mark the beak and upper wings and two appear on the tail. Tenderly clutched in his talons, six stars glow in the form of glorious Ganymede.[9]

For his part, Ganymede, the son of King Tros—the eponymous founder of Troy—also appears in his own constellation because of his steadfast service to the gods. Here he is known to the ages as the WATER BEARER. Every night, from his right hand he pours a sparkling libation marked by a shimmering stream of stars.

THE ALTAR OF HEAVEN

The campaign against Cronos and the Titans dragged into months, and months turned into years, but the war only worsened. Heaven and Earth quaked in fury, the seas slammed fiercely against the shores, and all of nature trembled.

To keep the army together in trying times, Zeus and the gods often affirmed their allegiance to one another by offering fragrant herbs on a sacred Incense Altar. In radiant robes they assembled, and in solemn procession they bore their blazing torches. One by one, they laid them on the altar's burning brazier, and swore undying devotion to their cause.

Thus armed—with fervor forged by oneness of mind—they sallied forth to destroy the foe. At last, ten tragic years of warfare ended. In dismal defeat, Cronos and the Titans descended into the dungeon of Tarturus, in the depths of the Earth.

That night, on the summit of shining Olympus the deities delighted in triumph. Nike—the goddess of victory—resplendent in starlight, blessed the moment by pouring pure libations. Then,

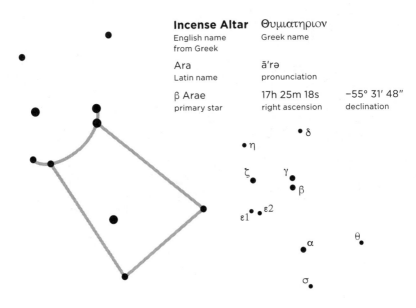

Incense Altar   Θυμιατηριον
English name   Greek name
from Greek

Ara   āˈrə
Latin name   pronunciation

β Arae   17h 25m 18s   −55° 31′ 48″
primary star   right ascension   declination

Nike (Victory) Pouring a Libation at an Altar. Painting on a
Lekythos (oil flask), c. 490 BC, from Attica. Attributed to the
Berlin Painter. Courtesy of the Harvard Art Museums/Arthur M.
Sackler Museum, loan from Estate of Donald Upham and Mrs.
Rosamond U. Hunter, 4.1908. Photo: Imaging Department ©
President and Fellows of Harvard College, Boston, Massachusetts.

Zeus placed the INCENSE ALTAR, with its rounded brazier and flying sparks, as a memorial in the sky. There it glows as a reminder to immortals and mortals alike of the strength of sacred devotion.[10] Now, all who see it in the sky are inspired to erect altars in holy places and pray for success in worthy endeavors. Now they affirm allegiances with "incense and reverent vows, and libations, and the savor of sacrifice."[11]

2

VIRTUE, LOVED AND LOST

CONSTELLATIONS
—
MAIDEN (Virgo)
ARCHER (Sagittarius)
SOUTHERN WREATH (Corona Australis)
CENTAUR (Centaurus)
WILD ANIMAL (Lupus)
TRIANGLE (Triangulum)
CROW (Corvus)
CRATER (Crater)

STARS
EAR OF GRAIN (Spica)
HERALD OF THE VINTAGE (Vindemiatrix)

Left: Jamieson Plate 18: Virgo

Zeus, the god of sky and storm, wielded the awesome power of lightning and thunder. Into battle he bore his jagged bolts and hurtled them with fury at his foes. With this weapon, he led the Olympian deities and faithful mortals to vanquish the Titan host. Now Zeus ruled unrivaled over earth and sky.

The victory of the gods brought happiness and harmony to the human race. People dwelled in peaceful repose with neighbors and with nature. They cherished the flora and fauna that flourished around them. They remained respectful—fair and just—in close accord with all creation. In return, they found their needs for food and shelter fully met.

They thrived according to the words of Hesiod, the poet and prophet, who said, "Those who do not turn aside from justice at all; their city blooms and the people in it flower. For them, Peace, the nurse of the young, is on the earth, and far-seeing Zeus never marks out painful war; nor does famine attend straight-judging men. . . . For these, the earth bears the means of life in abundance."[1]

ASTRAEA, THE STARRY GODDESS

During this golden era, Astraea—the starry goddess of purity, peace, and benevolence—lived her immortal life among the child-like mortals she loved.[2] From ancient lineage she descended, as the luminous daughter of Astraeus and Eos, the god of Dusk and the goddess of Dawn.

Astraeus and Eos also brought forth other "shining stars with which the sky is crowned." Among them is Phosphoros—the morning star—who rises radiant above the eastern horizon, heralding the approach of her mother, Dawn. Their other children include the buffeting breezes: Boreas, the north wind; Notos, the south wind; Zephyros, the west wind; and Euros, the east wind.[3] Unlike their serene and stately sisters, the blustering brothers roam, ever restless—rushing about, pushing against one another to dominate the direction of the wind. Rarely do they rest for long in quiet and calm.

But Astraea's manner is always soft and sweet—the pleasant personification of peace. In the golden era, she found the simple innocence of humans charming and endearing. She often was seen among village elders, "ever urging on them judgments kinder to the people." And peace prevailed, "for not yet in that age had men knowledge of hateful strife, or carping contention, or din of battle, but a simple life they lived. . . . The oxen and the plow . . . abundantly supplied their every need."[4]

As a heavenly constellation, Astraea appears as a MAIDEN in virtuous youth. Beautiful and benevolent, she carries a bright star named EAR OF GRAIN in her left hand as an emblem of agrarian

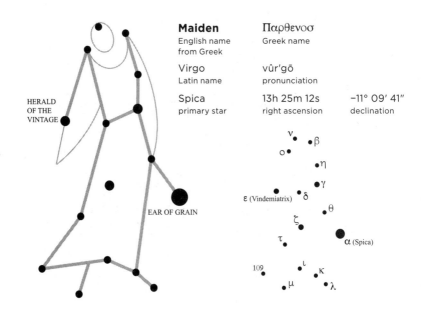

Maiden	Παρθενοσ	
English name from Greek	Greek name	
Virgo	vûr'gō	
Latin name	pronunciation	
Spica	13h 25m 12s	−11° 09′ 41″
primary star	right ascension	declination

HERALD OF THE VINTAGE

EAR OF GRAIN

ν
β
ο
η
γ
ε (Vindemiatrix) δ
θ
ζ
τ
α (Spica)
109
ι
κ
μ
λ

bounty. On her right wing rests a beam of light called HERALD OF THE VINTAGE. This is the star that annually rises on the eastern horizon at dawn as a sign for farmers to hasten to the vineyards and harvest the ripening grapes.[5] In her right hand she raises a palm leaf as a symbol of peace on Earth. Adorned with these—the greatest of earthly gifts—Astraea, the Maiden, shines above as the everlasting celestial sign of divine love for humankind.

In those golden days, other gods and goddesses also granted blessings to earthly beings. Even the centaurs—those galloping, gamboling creatures, half-man and half-horse—received divine gifts from immortal gods. Much like mankind, centaurs range in character from worthless to worthy. But two among them are always held in highest esteem—as high as any human.

THE CENTAUR OF THE MUSES

One of these centaurs is Crotos, the son of Eupheme. Eupheme, a gentle, mild-mannered centaur, served as nanny to the nine maiden Muses—the daughters of Zeus—who reside on the heights of Mount Helicon. On the day that Eupheme gave birth to Crotos, in a shady meadow by a mountain spring, the nine divine sisters helped with the delivery.

The girls giggled in mirthful delight as the tiny centaur, newly born, rose to his feet and wobbled forth on four spindly legs. In time, he found his balance and began to buck and play.

The sisters fell fast in love with the little fellow right away. Adopting him as a brother, they doted on Crotos from that day forward. Through the years, the young centaur benefitted from the blessings of their constant encouragement and kind instruc-

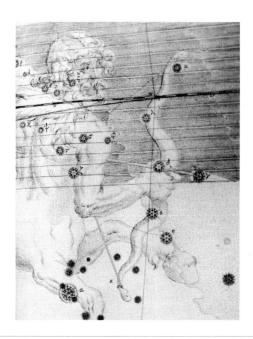

Sagittarius depicted in the *Uranometria* of Johann Bayer, 1603. Courtesy of the Museum of History and Science Library, Oxford, England. Photo by author.

tion. Unlike most centaurs, he enjoyed a civilized, cultivated life in their care.

Crotos soon advanced in childhood skills. He reveled in racing with woodland friends. He thrilled at clambering atop rocky pinnacles and peaks. He excelled in notching a feathered arrow in the string of his recurved bow, drawing it back, and sending it swiftly in flight to knock acorns or leaves off holm oak trees.

As Crotos grew older, the Muses immersed him in learning. His knowledge of the universe broadened—in time and space—through the study of history and astronomy, as well as other sciences. With the talented Muses as tutors, his love of the arts also flourished. Their divine duty, after all, was to share the arts of

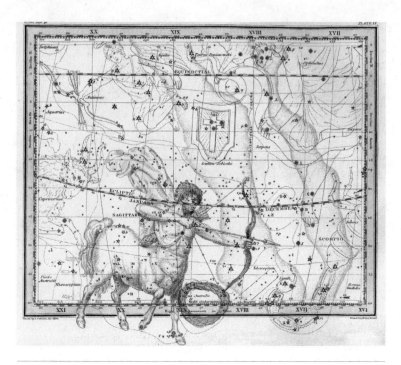

Jamieson Plate 20: Sagittarius, Corona Australis

poetry, music, dance, and drama, in addition to serving as patron-goddesses of history and astronomy.[6]

With their inspiration, Crotos bloomed as a brilliant and bold musician. He developed the method of emphatically marking the rhythm of songs and stories with claps of hands and stamps of feet. He also showed how to follow finales with grateful applause, formed by faster clapping of the hands.

For these innovations, the nine divine maidens lavished endless praise. The rhythmic beats brought their epic tales and music to life for listeners; and applause allowed an audience to respond in joyful ovations. Often, when the Muses performed their livelier songs or stories, Crotos could not contain himself but joined with a spirited dance. Taking his olive-branch wreath from his head,

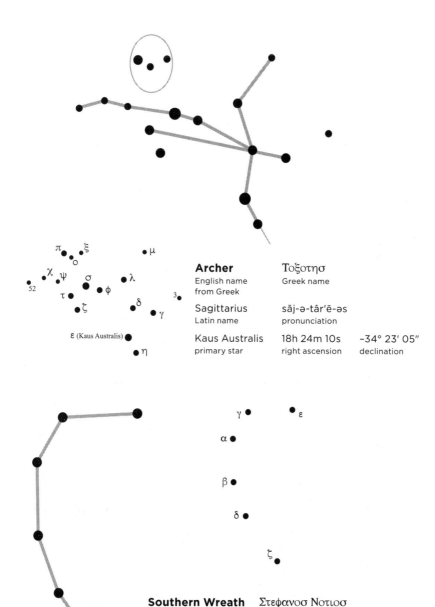

Archer
English name
from Greek

Τοξοτησ
Greek name

Sagittarius
Latin name

săj-ə-târ′ē-əs
pronunciation

Kaus Australis
primary star

18h 24m 10s
right ascension

−34° 23′ 05″
declination

Southern Wreath
English name
from Greek

Στεφανοσ Νοτιοσ
Greek name

Corona Australis
Latin name

kə-rō′nə ôs-trā′lĭs
pronunciation

α Coronae Australis
primary star

19h 09m 28s
right ascension

−37° 54′ 16″
declination

VIRTUE, LOVED AND LOST

he tossed it on the ground. Then, he merrily pawed and pranced around it, and jumped with leaps and bounds.[7]

The Muses and the two centaurs—mother and son—lived joyful lives together and became like family. At the same time, Crotos' pious devotion to the nine divine sisters and their father—Zeus—grew stronger through the years. But, alas, each passing year also saw the centaurs growing old and gray. At last, the immortal Muses—forever young—had to watch with heavy hearts as Eupheme and Crotos fell into feeble age and passed away.

When Crotos breathed his last, the grieving girls asked Zeus to make a place for him—their fondest friend—in heaven. And so, among the stars, he continues to roam as though rambling through the forest. He carries a recurved bow clutched tight in his left hand and drawn back by the right. His wreath of olive leaves lies on the ground before him. Crotos' constellation is called the ARCHER and his wreath is the SOUTHERN WREATH, because of its location, low in the sky to the south beside his prancing hooves.

CHIRON, THE WISE

Another centaur—Chiron—achieved the same acclaim for his piety and compassion. Like Crotos, he shared none of the savage traits that characterized so many centaurs. Instead, he remained steadfast in mind and spirit.

Chiron surpassed all others in the knowledge of botany and medicine, and excelled in celestial lore and music. He shared this wealth of wisdom by mentoring several students, who rose as lofty pillars of learning and leadership. One of these students was Jason.

When Jason was only an infant, his wicked uncle—the king of Thessaly—plotted to kill him in his cradle to prevent, one day, his

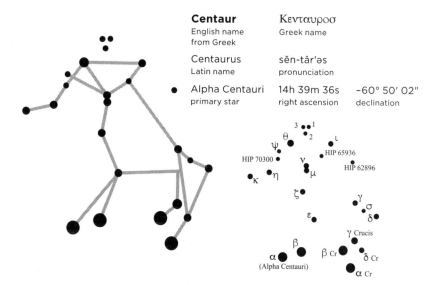

Centaur
English name
from Greek

Κενταυροσ
Greek name

Centaurus
Latin name

sěn-târ'əs
pronunciation

Alpha Centauri
primary star

14h 39m 36s
right ascension

−60° 50′ 02″
declination

3 ••¹
θ •²
ψ •   ι
HIP 70300 •   • HIP 65936
ν
κ •η   μ   HIP 62896
ζ•
ε•   γ
•σ
δ•
γ Crucis
β   •
α ●●   β Cr ●   δ Cr
(Alpha Centauri)   α Cr

possible bid for the throne. Chiron pitied the innocent child and hid him in his cavern home on the heights of Mount Pelion. There he raised him as a son, imparting his knowledge and lore.

By Jason's side, the child Asclepius—another pupil of Chiron— practiced the healing arts. Asclepius excelled as a promising protégé, and Chiron freely bequeathed to him a command of the curative properties of plants. When it came time to set aside the studies and go outside to play, the boys—Jason and Asclepius— loved to wander the woods together.

Through towering fir trees and shady glades, they climbed to the heights of Mount Pelion. The farther they roamed, the less they feared the deep, dreary forest that surrounded their cavern home. Along the way, they watched for bees that would lead them to wild honey trees. And they learned to recognize useful plants to add to Chiron's herbarium.

Some years later, when the boys at last left their childhood cave as men, Asclepius chose to journey with Jason and his adventurous band of Argonauts. As a worthy member of the crew, Asclepius

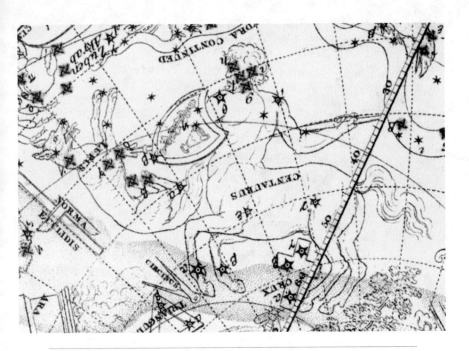

Jamieson Plate 28, detail: Centaurus, Lupus

served as ship's physician on the distant voyage to Colchis. At Chiron's behest, Jason also invited Orpheus, the famed musician, to join the quest.

Chiron, the kindly centaur, supported the Argonauts' valiant endeavor from start to end. He offered prayers as the men prepared to sail their ship—the *Argo*—into unfamiliar waters. Then, when the crew loosened the lines and heaved their ship into the surf, Chiron descended to the pebble shore below Mount Pelion. With an impassioned plea to the gods for the safety of his sons—hidden in mumbles beneath his breath—he waved farewell, and, with a smile, wished them a swift return.[8] What became of them, we will later learn.

Over the years, rumors of Chiron's wisdom spread far and wide, and many men sought his sage advice. Often they searched

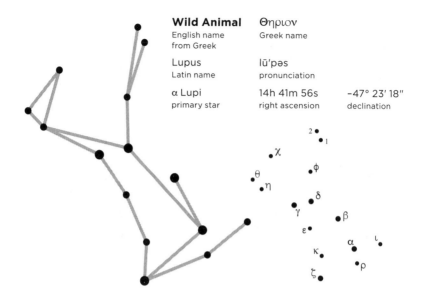

Wild Animal
English name
from Greek

Θηρίον
Greek name

Lupus
Latin name

lū'pəs
pronunciation

α Lupi
primary star

14h 41m 56s
right ascension

−47° 23' 18"
declination

the forested slopes to find the shady pathway to his cave. Heracles (Hercules)—the son of Zeus—knew the trail quite well and frequently came to request the centaur's counsel.

Once, as the two enjoyed a lively conversation in Chiron's cave, a poisoned arrow—tipped with the venomous blood of Hydra—suddenly slipped from Heracles' quiver and punctured Chiron's lower leg. Chiron collapsed and fell into a coma. In horror, the mighty hero, suddenly helpless, held his centaur friend and watched as he closed his eyes and slipped away.

As earthbound mortals mourned the tragic loss, the gods gathered and agreed to honor Chiron in the eternal heaven. His constellation—simply called the CENTAUR—stands tall on four legs and walks with a stately gait through the celestial sphere. In his hand he holds a WILD ANIMAL—a savage stalker of his forest home, which most perceive in the shimmering sky as a wolf. The stars show Chiron clasping the hind feet of the beast in his right hand. In the left, he carries a thyrsus (a staff adorned with leafy vines and pine cones) in devotion to Dionysus, the god who likewise loves to wander the wooded mountains.

Crotos and Chiron both gained the timeless praise of immortals and men. Both embodied the qualities that Hellenes most admire: skill at arms and in the arts; valor devoid of vanity; an adventurous and enlightened spirit; knowledge and wisdom; compassion and piety. They portrayed the essence of *arete*. Appropriately, the two appear in the sky as they approach, in reverence, the Incense Altar of the gods. In proper display of their personalities, Chiron is shown in solemn procession and Crotos in joyful dance.

THE END OF A GOLDEN ERA

Crotos and Chiron lived their lives during glorious times for mortals. But golden ages must come to an end, and centuries sometimes lapse before they appear again. Mortals, male and female, young and old, alas, fell into the folly of conceit. They came to consider themselves the center of all creation. In blind lust to rule the world, they soon lost sight of their oneness with nature and the spiritual essence around them.

The beauty of Astraea's divine blessings—of virtue, peace, and plenty—dimmed in a fog of ingratitude. At last, the goddess found herself alone—ignored as if invisible—without the laughter and love of her earthly children. Weeping at their woeful deeds, and the dreadful consequence sure to come, she fell into deep despair. Sobbing in soulful lament, she left the world behind and ascended into heaven with a broken heart.

Without her peaceful presence, human greed for power and possession brought the horror of hatred and war among them.

And so began the sordid age of the restless workers of bronze, "who were the first to forge the sword . . . and the first to eat of the flesh of the plowing-ox."[9] Without a divine beacon and a moral ethos as their guide, they turned on each other like wolves—"and they slaughtered with the bronze."[10]

The kingdoms of Hellas and Asia now warred with one another from one side of the Aegean Sea to the other. Some blame the Phoenicians or Persians of Asia for instigating these wars of old. Others accuse the Hellenes. Herodotus, the historian, recorded the Persian account that said a crew of Phoenician sailors arrived at Argos one day. Here, they planned to barter their Asian goods for olive oil and purple wine, and the painted pottery of Hellas. After several days of successful trade, the sailors prepared to depart.

As they rigged the ship and coiled the lines, they suddenly grabbed several girls of Argos who had come to the docks to shop. Before there was time to react, the Phoenician crew caught a freshening breeze—a favorable wind from the west—and fled far over the eastern horizon. One of the mournful, captive maidens was none other than Io, the princess of Argos. Her father, the king, was furious and demanded revenge for his daughter's abduction.

This was the Persian story. But Hellenic traditions tell us that Io was captivated by Zeus. The covetous god made love to the maiden, then transformed her into a snow-white heifer to hide and protect her from Hera, his jealous wife. But Hera knew the ways of Zeus and soon saw through the disguise. In a fit of fury, she sent a gadfly to chase the girl far from her native home.

With the biting fly in hot pursuit, Io ran bucking and kicking along the shore of the Ionian Sea, which bears her name. From there, she sped eastward and swam the Bosporus—or *Cow Ford*—which carries her name as well. Finally, she fled far to the south, leaving the gadfly behind, and found a safe haven in Egypt.

Located at last in the land of the Nile, Zeus restored Io to her former youthful figure. Or so goes the Hellenic tale.

According to the Persian account, after Io's abduction from Argos, a Hellenic merchant ship crossed the Aegean Sea and moored at the Phoenician port of Tyre, on the Asian coast. Upon completing their commerce, the crew abducted Europa, a princess of Tyre, in retaliation for Io. Not long later, another crew of Hellenic sailors, trading at the eastern city of Aea in Colchis, absconded with another Asian princess, named Medea. Of course, Hellenic versions vary, as we shall see.

The Persians claimed that this double abduction of eastern princesses prompted Paris—the prince of the Asian city of Troy—to capture Helen, the queen of Sparta in Hellas. With beautiful Helen whisked away to Troy, the kings of Hellas reacted in wrath and rage. Together they launched a thousand ships filled with their fiercest warriors. Sailing eastward across the Aegean Sea, they planned to retrieve the Spartan queen and to plunder the prosperous city of Troy.

For ten bitter, tempestuous years they waged their war. Blood from countless battles stained the sand from the city walls to the Hellene tents and ships on the shore. The furious fighting "sent down to Hades many valiant souls of warriors, and made the men themselves [the lifeless bodies on the beach] to be the spoil for dogs and birds of every kind."[11] In the end, proud and prosperous Troy lay in wreck and ruin—smoldering and choking in ash and smoke.

Paris, Prince of Troy, had acted like the greedy dog in Aesop's fable. The dog had all the meat he could eat clutched firmly in his mouth; but he lost it all when he opened his jaws to grab for even more.[12] Paris had the royal wealth of Troy at his disposal; but, by taking Helen, he lost his life and the lives of many thousands.

As a legacy, he left his ancestral land in desolation. Still worse, the Trojan War sparked eight centuries of conflict between the empires of Asia and Hellas. Hundreds of thousands of mortals—belligerent and blameless alike—died on both sides.[13]

The same disdain for humans and gods that kindled the Trojan War also caused the armies of Hellas to further provoke the divine wrath. Odysseus and his Ithacan army had joined their Hellenic countrymen in the ten-year quest to conquer Troy. In the final months of fighting, Odysseus—clever and cunning as ever—devised the victory by building a massive wooden horse. Leaving it alone to tower above the sandy shore, as if an offering to Poseidon, the Hellenes launched their fleet and headed for home.

When the Trojans discovered their hasty retreat and the huge horse looming above the beach, they shouted aloud in triumph. Quickly, they hauled the wooden hulk inside the city's impregnable walls and placed it as a gift in Poseidon's temple. After a day of drunken debauchery, the city slept soundly that night.

Unknown to them, the lifeless horse bristled with bloodthirsty Hellenes—crowded into every corner within. Then, when all was dark and quiet, the waiting warriors stealthily slipped to the ground and opened the city gate for the returning Hellenic army. Ten years of bitter rage at last became unleashed as the Hellenes ruthlessly slaughtered Trojans—men and women, young and old. For this massacre, and for the mock offering of the horse, Poseidon promised to punish the mastermind—Odysseus—when he sailed for home on the stormy sea.

Thus it happened that Odysseus won the war only to be "driven far astray after he had sacked the sacred citadel of Troy." While trying to return to his island kingdom of Ithaca, Poseidon condemned him to wander the merciless waters for ten more years. All the while, Odysseus endured endless hardships and close encounters with death. Many were "the woes he suffered in his heart upon the sea, seeking to win his own life and the return of his comrades" to Ithaca—where his long-suffering wife, Penelope, patiently waited.[14]

THE BEAUTIFUL ISLAND OF SORROW

Odysseus and his starving men, while lost and adrift on the angry sea, made sudden landfall on the gentle, sloping shore of a sunlit island. This paradise in the middle of the Mediterranean bore the name Thrinacia, because of its strange triangular shape. Centuries later, sailors from Hellas would find the island rising, beautiful to behold above the surging sea, and rename it Sicily.

At one time, the island offered a favorite haven for immortal gods.[15] Here, Apollo kept his handsome herd of sacred cattle, safe and secluded. Like Apollo, Demeter—the goddess of farming and harvest and protector of green and flowering plants—loved the island for its lush verdure and pastoral charm. She even called on Zeus to honor the pleasant isle in heaven, where it shines as a simple TRIANGLE bound by three stars.[16]

But sorrow soon followed on Thrinacia. Demeter's dreamy daughter, Persephone, loved to wander with her maiden friends through the island's dewy meadows, laughing softly and gathering fragrant flowers. Her soothing voice and glowing countenance cap-

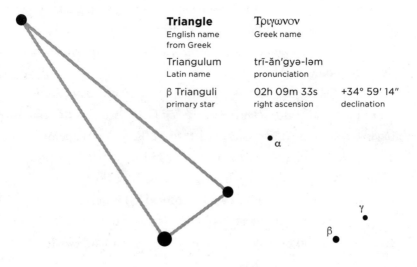

Triangle
English name
from Greek

Τριγωνον
Greek name

Triangulum
Latin name

trī-ăn′gyə-ləm
pronunciation

β Trianguli
primary star

02h 09m 33s
right ascension

+34° 59′ 14″
declination

Demeter (left) and Persephone (right) with Triptolemus. Marble relief by Phidias, c. 440 BC. Courtesy of the Archeological Museum, Eleusis, Greece. Photo by author.

tured the heart of Pluto—the god of the underworld. Pluto rarely witnessed anything other than darkness and death. But from the depths of a shadowy chasm, he often watched the girl's delightful form as she glided on dainty feet through the glistening fields.

One sunny day, Persephone spied a colorful blossom dancing in the breeze near a deep crevice. Lured by its beauty, she left her maidens and hastened to the beckoning flower. As she approached, Pluto burst from below and carried the frightened girl to his dreadful domain to make her his wife.

Demeter, on learning of her daughter's abduction, wailed in heartfelt anguish day and night. Again and again, she pleaded with Zeus to punish Pluto and return her daughter to her tender embrace. Zeus, in time, intervened. But he could only compel his brother, Pluto, to allow Persephone to spend half of each year with her mother in the delightful world she loved. Thus, for six months, as Demeter grieves for her daughter, the cold darkness of winter descends upon the earth. But with the return of Persephone to her mother's arms each spring the earth abounds in flowering plants and the songs of birds.

Tragedy struck again on Thrinacia when Odysseus' famished men ignored their king's strict command and killed and ate some of Apollo's sacred cattle. For this sacrilege against the gods, and for rousing the wrath of Apollo, Odysseus and his crew ran headlong into a fearsome storm as soon as they sailed from the peaceful island. As the crashing waves cracked the beams and shattered the ship into splinters, every member of the crew floundered and drowned in the dark and swirling waters.

Only Odysseus, who abhorred the actions of his men and prayed for divine forgiveness, was spared. But still he suffered the woe of being the sole survivor of the proud Ithacan army that had sailed for Troy so many years before. As further punishment, the deities delayed his homeward return for several years more.

THE GREEDY CROW OF APOLLO

Men were not the only mortals to fail in their devotion to the gods, and suffer the result of divine wrath. Even a few of the animals most trusted by the Olympians fell out of favor for putting their own interests first. Such was the case of the Crow—Apollo's favorite bird and close companion.

One day, Apollo called on the Crow to bring him water from a sparkling pool on Mount Olympus so he could offer a holy libation. In those days, the immortals often affirmed their allegiance to one

Apollo holding his Lyre and pouring a libation, with his Crow in attendance. Vase painting, c. 470 BC. Courtesy of the Archeological Museum, Delphi, Greece. Photo by author.

another by pouring pure drops of water as a sign of esteem. This was long before wine was invented and used for the same purpose.

The Crow swiftly obeyed Apollo and seized a gleaming Crater—the goblet of the gods—in his talons. Searching high and low, the Crow scanned the slopes for a spring that offered a draft of water worthy of a divine libation. Soon he arrived on the shady bank of a shining pool. As he began to fill the Crater with crystal clear water, his gaze fell on a voluptuous fig tree bending low with the weight of its fruits.

Hungrily, greedily, the Crow hopped from limb to limb, waiting

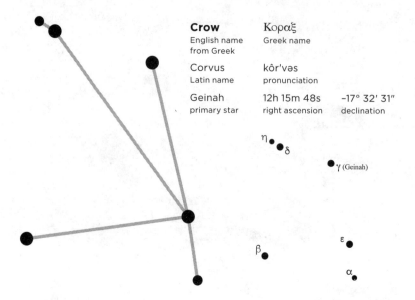

Crow
English name
from Greek

Κόραξ
Greek name

Corvus
Latin name

kôr'vəs
pronunciation

Geinah
primary star

12h 15m 48s
right ascension

–17° 32' 31"
declination

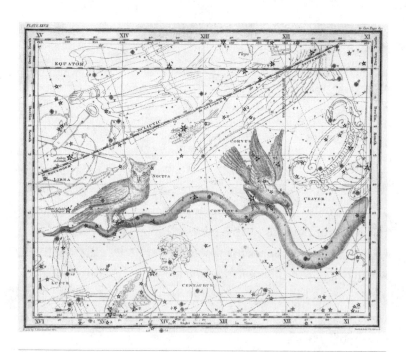

Jamieson Plate 27: Corvus, Hydra

ANCIENT SKIES

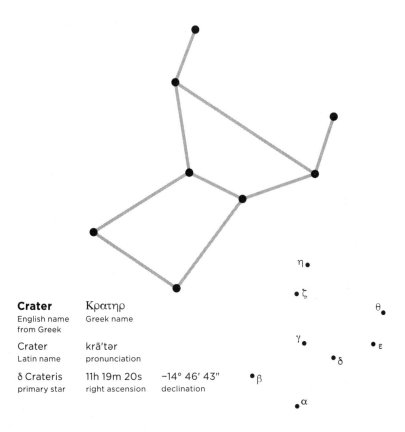

Crater
English name
from Greek

Κρατηρ
Greek name

Crater
Latin name

krā'tər
pronunciation

δ Crateris
primary star

11h 19m 20s
right ascension

−14° 46' 43"
declination

for the figs to fully form so he could fill his belly before returning to Apollo. One day turned into two, and two into three. Finally, the fruit was ready—plump and ripe—and the Crow gorged himself with delight.

With his appetite now satisfied, he began to ponder Apollo's reaction to his impulsive delay. After a moment's hesitation, he grabbed the Crater in one claw, and in the other a wondrous water snake that guarded the sacred spring. Then he flew fast and true— with all possible speed—to Apollo.

His head remained bowed and his eyes cast down as he tried to explain his lengthy stay at the spring. With many reluctant stops and starts, he told a tale—how the huge snake had drunk the water dry, day after day, and kept him from filling the goblet. But Apollo could not be so easily fooled. He punished the Crow and all of his kind by making them go without water for days at a time, with dry and cracking throats, and rasping calls. While they soar above the fields and forests, they caw aloud for a soothing draft to quench their nagging thirst.

Furthermore, as a warning to those who choose to put their personal greed before the gods, Apollo placed the CROW in the nighttime sky. There, the boisterous bird, with wings held high, pecks at a serpent—much like the snake at the spring. Just out of reach of the Crow is the starry CRATER, holding a cool draft of crystal clear water.[17]

3

WOEFUL DEEDS OF THE WANTON GODS

CONSTELLATIONS

BEAR (Ursa Major)

BOOTES (Bootes)

LITTLE BEAR (Ursa Minor)

BIRD (Cygnus)

BULL (Taurus)

NORTHERN WREATH (Corona Borealis)

ASTERISMS:

PLEIADES (Pleiades)

HYADES (Hyades)

STAR

BEAR GUARD (Arcturus)

As mortals came to disdain their fellow beings and divine deities, selfish struggle and strife drove them to war among each other. The gods—indignant—fumed in anger. Even Astraea, who had loved humans as innocent children, walked away in sorrow. Zeus himself, who had chuckled in mirth at the childlike folly of men, now turned bitter toward them and became capricious and cruel. Innocent people suffered along with the rest, as the stories of Callisto, Arcas, and Phoenice attest.

CALLISTO, THE MOUNTAIN MAIDEN

In the distant past, the maiden Callisto dwelled among the dark forests and jagged crags of Arcadia. Hermes had been born in this wild and rugged country, in a rocky cave on Mount Kyllene. And Pan—"the goat-footed, two-horned rowdy," the patron of savage beasts and shepherds' flocks—still haunted the deep hollows and dizzying heights.[1]

The mysterious mountains enchanted Callisto. She often wandered, beneath the moon, through meadows glowing in subtle light and forests shrouded in shadows. In euphoric suspense, she sometimes stopped and stood silent and still in the darkness. She listened intently and peered through the gloom for a glimpse of the wild and wary creatures that prowled around her.

Callisto followed, in firm devotion, the divine example of Artemis—goddess of the hunt and matron of the moon. Artemis, too, delighted in spending days and nights softly stalking through somber forests in search of game. The goddess—a virgin—also championed female chastity, and Callisto had sworn to abide by that code. But the dictates of fate were not for her to decide.

One morning, while Zeus was seeking solace in the dusky

woods, he spied Callisto from a distance—gliding silently from tree to tree and fully focused on the hunt. The maiden remained unaware of his presence as he followed her through the forest, until he met her face to face. Unable to fend against his advances, she conceived a child and gave birth the following spring to a son.

Forlorn, and afraid of Artemis' anger at the conception and birth, Callisto bundled her baby and held him close—concealed in her cape. But the virgin goddess heard the crying child and found that the girl had not remained chaste. In misplaced rage, she rashly blamed Callisto for the indiscretion.

As punishment, Artemis snatched Callisto's son away and transformed her from a graceful youth to a blundering bear. Her tender voice became a grumble and growl. Her silky hair turned course and matted, and covered her body from head to foot. Condemned to ramble the woods on four feet, as a brutish beast, Callisto was constantly hunted by her own countrymen. The barking of dogs and shouts of the chase that had filled her with thrills and excitement now brought fear and foreboding.

Many loathsome, lonely years passed, and Callisto's infant son— Arcas—grew to be a sturdy youth. Raised by a rustic goatherd in the rugged heights of Arcadia, Arcas earned his keep with a drove of goats and sheep, and the meat he brought home from the hunt. One calm morning, while his flock lazily grazed beside a trickling spring, he spotted a she-bear on the shadowy fringe of the forest.

With a leap and a yell, he gave chase, unaware that the bruin was his own woeful mother. All that day, he pursued the frantic beast through brambles and brush, down rocky ravines, and across cold mountain streams. At last, he wore her down. As she lay exhausted before him, her head on the ground, with pleading eyes turned upward, he held high his club to deliver the fatal blow.

The loud shouts of the hunter and the lamentations of the bear caught the attention of far-seeing Zeus. As he watched the tragic drama unfolding below, he felt sudden pity for Callisto and Arcas—his long-lost lover and child. In haste, he forestalled the evil fate that threatened to make one the victim of her son and the other the murderer of his mother. There, on the leafy forest floor, with a word from Zeus, the two suddenly knew their true relation, and, overawed, rejoiced in the reunion.

From that day forward, Callisto—who knew every haunt and hollow—led the way down hidden trails as she and Arcas wandered far and wide through the forest. Arcas always followed close behind, faithfully protecting his mother. Finally, to offer her safety and solace forever, Zeus placed her in the highest heaven, where she shines as the constellation Arctos—the BEAR.[2] Here she circles the north celestial pole in peace, with none to fear. And yet she keeps a wary eye on the stars that mark the hunter named Orion.

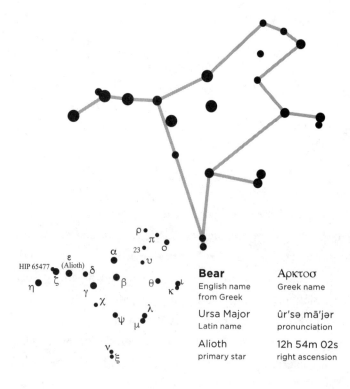

Bear
English name
from Greek

Αρκτοσ
Greek name

Ursa Major
Latin name

ûr′sə mā′jər
pronunciation

Alioth
primary star

12h 54m 02s
right ascension

+55° 57′ 35″
declination

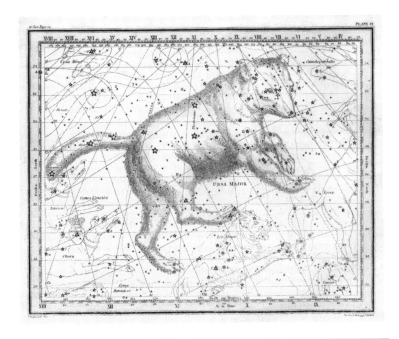

Jamieson Plate 6: Ursa Major

Arcas came to rule the rugged mountain kingdom, which was later named Arcadia in his honor. He lived a long and happy life in his forest home while shepherding his people well, and forever revering his lustrous mother above. When he came at last to the end of his days, Zeus set him in the cosmos as the Bear's protector, and called him BOOTES—the Shouter.[3] His brightest star is Arcturus—the BEAR GUARD—so named because of his constant watch over his mother's constellation. With a shepherd's staff in the right hand and his left hand reaching fondly toward her, he follows close behind in the northern sky.

When Zeus intervened to reunite the mother and son, Artemis at last learned the truth—that Callisto conceived the child through no fault of her own. In deep remorse for her unfair judgment and brutal reprisal, the goddess bowed her head in

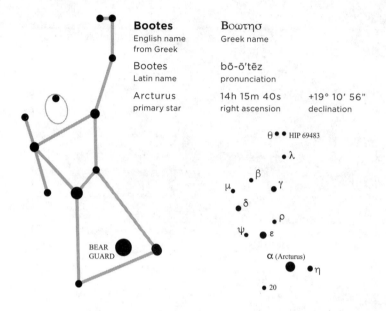

Bootes
English name
from Greek

Βοωτησ
Greek name

Bootes
Latin name

bō-ō'tēz
pronunciation

Arcturus
primary star

14h 15m 40s
right ascension

+19° 10' 56"
declination

θ ● ● HIP 69483

● λ

β
●

μ ●   ● γ

● δ

ρ
●

ψ ●   ● ε

α (Arcturus)
●   ● η

● 20

BEAR
GUARD

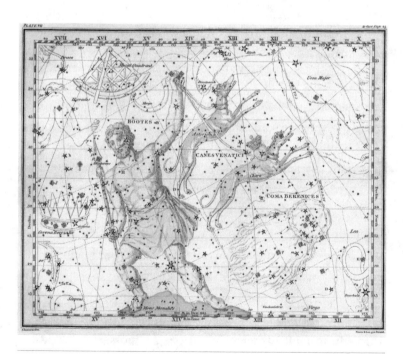

Jamieson Plate 7: Bootes

shame. Too late to make amends, she vowed to never take that path again.

In years to come, Zeus, the wanton god, lustily followed Phoenice—another faithful admirer of Artemis. Phoenice—a daughter of the Phoenician race—was much like Callisto in manner and morals. No sooner had Artemis detected Zeus's threat than she took fast action to protect the innocent girl. Reacting in pity, rather than rage, the goddess changed Phoenice's feminine form into that of a bear before any further harm could come.

But Phoenice was never forced to wander the woods in fear, like Callisto. Instead, Artemis set her securely in the sky as a constellation, in close company with her larger companion—the Bear. To distinguish the two, Phoenice is called the LITTLE BEAR. At the apex of heaven, she stalks swiftly forward through the night in a rapid circle, above the back of the Bear. The two stand out from earthly bruins because of their long and splendid, starry tails. The Little Bear's tail curves upward, so she sometimes carries the nickname Cynosura, or "dog-tail."[4]

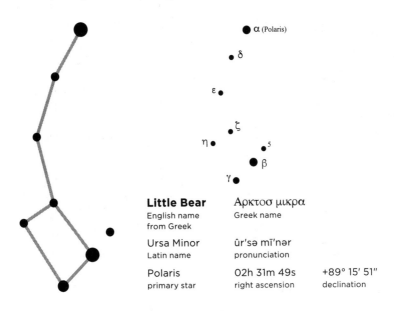

Little Bear	Αρκτοσ μικρα	
English name from Greek	Greek name	
Ursa Minor	ûr′sə mī′nər	
Latin name	pronunciation	
Polaris	02h 31m 49s	+89° 15′ 51″
primary star	right ascension	declination

THE SEDUCTIVE SWAN

Zeus continued to pursue mortal women. To seduce them, he sometimes dazzled their senses and captured their affections by appearing in the form of a magnificent animal. As a spectacular, snow-white swan, he captivated the love of Leda of Sparta. Soon she bore him twin sons—Castor and Polydeuces—and daughters, Helen and Clytemnestra.

The two boys later emerged as heroes of Hellas and journeyed with Jason and the Argonauts. Helen gained renown for her unrivaled beauty, and married Menelaus—the king of Sparta. Her abduction to Troy sparked the Hellenic conquest of that prosperous and powerful city. Helen's sister, Clytemnestra, wed the brother of Menelaus—King Agamemnon of Mycenae—who commanded the seaborne invasion.

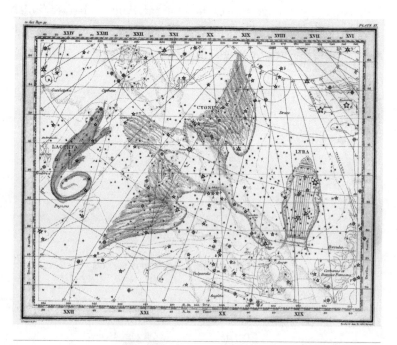

Jamieson Plate 11: Cygnus, Lyra

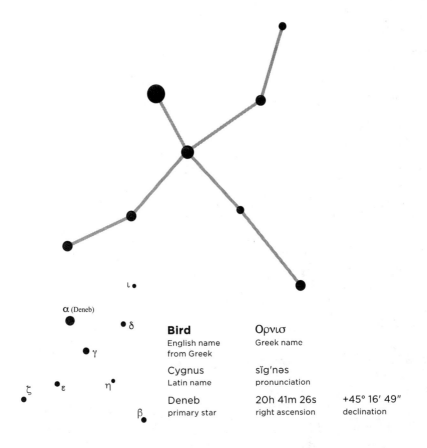

Bird
English name
from Greek

Ορνισ
Greek name

Cygnus
Latin name

sĭg'nəs
pronunciation

Deneb
primary star

20h 41m 26s
right ascension

+45° 16' 49"
declination

Zeus celebrated his liaison with Leda by placing a constellation, called the BIRD, in the sky in the shape of a swan. The swan shines bright among the dense assembly of stars in the glowing Milky Way. There, he appears as if "wreathed in mist" and soaring "like a bird in joyous flight."[5]

EUROPA AND THE BULL

Zeus assumed the shape of another stunning creature—a snow-white bull—to win the affections of the maiden Europa. From

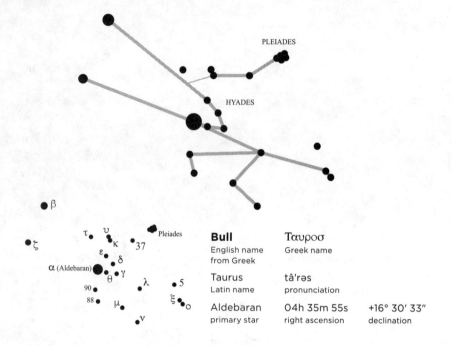

PLEIADES

HYADES

β

τ υ
ζ   κ   37   Pleiades
    ε   δ
α (Aldebaran)
        θ   γ
90   λ   5
88   μ
                ξ   o
        ν

Bull
English name
from Greek

Ταυροσ
Greek name

Taurus
Latin name

tâ'rəs
pronunciation

Aldebaran
primary star

04h 35m 55s
right ascension

+16° 30′ 33″
declination

childhood, the cheerful Europa had always loved the coastline of her native Phoenician home, much as Callisto adored the forested mountains of her land of birth. Europa often wandered along the Asian shore, laughing at the splashing surf and savoring the salty breeze. She watched for delicate seashells and other treasures dropped at her dainty feet by the benevolent sea.

One warm day, as Europa strolled along the balmy Mediterranean beach, she froze in her footsteps—entranced by the sight of a beautiful bull, grazing the seagrass near the shore. Approaching slowly, she petted the blond, curly locks between his horns. She picked pretty flowers and fashioned a garland to adorn his head. Finally, she grasped his supple mane and slipped atop his broad back for a wade in the rolling surf.

Suddenly, the docile bull sprang to life and lunged into the waves. Swiftly he swam, rapidly plunging through the waters

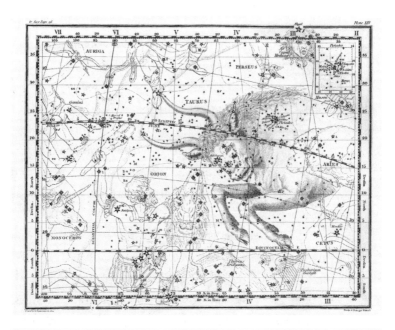

Jamieson Plate 14: Taurus, Orion

toward the west—sweeping Europa out to sea. As she desperately clung to his curly mane, she glanced behind and watched in dismay as her homeland dwindled and finally disappeared.

The island of Cyprus soon loomed ahead. But the bull continued to batter the waves, bounding toward the setting sun. At dusk, the island of Crete came into view. As they approached, it seemed to swell on the horizon. Here, at last, the bull with his maiden prize waded ashore to rest on the sandy beach.

Europa collapsed near the crashing surf, sobbing in woe and foreboding, and homesick for her family. But the charming island enticed her to wander its pleasant shores. By and by, she found that graceful beaches surrounded the elegant land in all directions. Her sorrow now gave way to glee as the shorebirds sang her a welcoming song, and a gentle sea breeze softened her tears.

Crete soon captured her heart and became her adopted home.

The following year, she brought forth a son and named the baby Minos. As a toddler, the child—the son of Zeus—traipsed hand-in-hand with his mother along every sunbaked beach of the island. And, like his mother, he treasured the gifts of the sea.

While still a youth, he hiked the hills alone and came to love the land—every cave and cliff, every mountain, meadow, and forest. Soon he knew all the native beings of the island, the plants, animals, and people, by name. So no one seemed surprised when, one day, Minos—the son of Europa and Zeus—reigned as king of Crete.

Zeus was pleased to see the happiness of the mother and son. In satisfaction, the powerful god—ever proud of himself—commemorated his cunning theft of Europa by placing the BULL in the nighttime sky. The majestic beast blazes among the stars, half-submerged and splaying his forelegs in the same manner that he swam the heaving sea.

The Bull is called a constellation "rich in maidens," because he bears several sisters on his back, just as he once bore Europa. The seven sisters, called the PLEIADES, appear as seven stars that ride high on his hump as an asterism. Their five half-sisters—the HYADES—shimmer as heavenly lights on his head.[6]

THESEUS AND ARIADNE

Minos ruled the people of Crete with fairness. But he fell into the common folly—far too often repeated—of showering his affection on one of his several children, to the detriment of all else. In fact, Minos adored his son, Androgeus, more than the rest of his family, his friends, and his entire kingdom.

Under his father's constant attention, with lavish expenses for coaches and training, the boy developed into a talented athlete. When Androgeus traveled to Athens and won the victory at the Panathenaic Games, all of Crete rejoiced beside the beaming father. But many Athenians fumed at the thought of a foreigner wearing the triumphal wreath of their city.

Several ruffians banded together in the dark and inflamed each other with angry words. Then, in a fit of fury, they followed Androgeus and crowded close to kill the youth. The people of Crete reeled when they heard the hated news. Minos locked himself in his palace—unseen for days—distraught beyond words and weary of the world's evil ways.

At last, his sorrow surrendered to rage. In brutal retaliation, Minos imposed a terrible tribute on Athens. The wrathful king

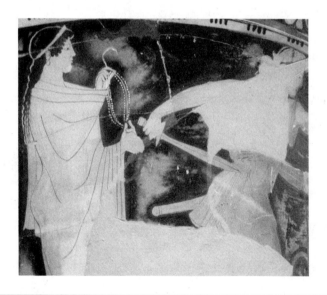

Ariadne offering a thread to Theseus. Vase painting, c. 480 BC. Courtesy of the National Archeological Museum, Athens, Greece. Photo by author.

forced the fair city to send seven of its sons and seven of its daughters as annual sacrifices to the man-eating Minotaur.

The monster, half-man and half-bull, dwelled in the deep, dark labyrinth—the twisted maze of Cretan caves. Alone in that dreary domain, the hideous, half-starved creature raged within his wretched world, bellowing day and night for another screaming victim to meet his miserable fate.

If the Athenians failed to pay the horrid tax of human flesh, Minos—the powerful King of Crete—promised to burn their city and level their land. For many years, Athens had no option but to pay the dreadful debt. As yet another year approached, the Athenian Councilmen, wringing their hands in anguish, tried to decide which of their fair children to send to their gruesome deaths.

At that moment, a brash young man named Theseus stepped boldly forward. To their speechless surprise, he proposed to journey to Crete as a would-be sacrifice. There he would battle the beast and win a cease to the grisly tax. The Councilmen opposed the plan in theory, but not enough to dissuade themselves of its merit. In the end, they accepted Theseus' offer.

Soon the Athenian ship, with its confident hero and crew, and thirteen tearful boys and girls, found its way across the wine-dark sea to moor on the coast of Crete. Theseus—tall and fiery— was first to set foot on the sandy shore. Immediately, without his knowing, he captured the eyes and heart of Ariadne—the daughter of Minos.

Without hesitation or need for further thought, the Cretan princess determined, then and there, to save the handsome youth from the Minotaur. As the curious crowd of gawkers departed and an opportunity arose, Ariadne sidled close beside him. Softly, she whispered hurried words of advice in his ear, and slipped him a simple ball of thread.

Thus armed, Theseus entered the labyrinth the following day

with confidence, while his companions trembled and wailed at the entrance. Leaving them there, he silently stole through the gloomy cavern. As he slipped warily forward, he unwound the spool along the path to provide a guide for escaping the twisted lair. Far down the dismal route he wandered, passing fly-infested heaps of putrid bones. Suddenly, he stumbled upon the foul fiend in the darkness. The battle raged, with the two powerful enemies beating and butting each other until they fell to the blood-soaked floor exhausted. Heaving and frothing, Theseus was first to regain his strength and catch his breath. With bare fists, he pummeled the Minotaur until the monster's eyes glazed over in death.

Now the hero made his slow escape, following a simple, indispensable string—his lifeline from the mind-numbing maze. As he staggered out of the somber gloom, he shaded his eyes against the bright and cheerful light of the sun. From somewhere near, the other Athenian youths, still shaking with fear, heard the frightening sound of approaching footsteps. But soon they stared in disbelief, then cried in overwhelming relief as their comrade came into view. Ariadne, having saved Theseus by a thread, rejoiced in silence from her hiding place in a nearby grove of trees.

Undetected by the palace guards, Theseus and his thirteen friends dashed toward the Athenian ship where the hopeful crew awaited. Beside them ran the Cretan princess. Her mind firmly fixed, Ariadne abandoned her father and country and sailed in haste with the hero, in hope of a new life in Athens.

At mid-voyage, while resting on the island of Naxos, the maiden—her body, mind, and emotions fully exhausted—fell into deepest slumber. For long hours she laid so silent and still that Theseus feared she had passed into peaceful death; every attempt to revive her failed. Finally, with his happy heart now broken, Theseus left her lying in sweet repose and resumed the long journey without her.

With judgment clouded by grief, he forgot the promise he had made to his father to hoist a white, billowing sail as the ship approached the Athenian shore. The sail was to serve as a signal that Theseus had found his way home, alive and well. Instead, as he slept, the unknowing crew sailed the ship close to the coast with the traditional black sail of mourning hoisted high on the mast.

Theseus' father had silently sat, day after day, with his eyes fixed on the horizon. At Poseidon's temple at Sounio, on a precipice above the sea, he watched and waited for his son's return. When in the distance he saw the dreaded canvas, billowing black above the sea, he knew it meant that Theseus and the thirteen youths had fallen to the ferocious Minotaur.

In sorrow and despair, he cast himself off the cliff, into the churning waters below. When Theseus stepped ashore, he heard the horrid news. At once, his victory evaporated—like a fresh drop of rain on a blazing summer day—whisked away by the double tragedy of losing his father and Ariadne.

Back on the island of Naxos, Ariadne finally awoke to find her loved one gone, and herself alone. The poor girl could not be consoled. In so short a span, she had lost her beloved hero, her family, her friends, and her homeland.

Her mournful sobs and deep distress attracted the attention of the god, Dionysus. In pity for the trembling maiden, he soothed her brow and stroked her hair, and soon fell fast in love. At last, she set her sorrow aside and vowed to go with him to the heights of Mount Olympus to be his bride.

There, the delightful wife of Dionysus became a favorite among the immortals. Many years later, when Ariadne passed away, all of Olympus mourned. Dionysus, in deepest sorrow, set his loved one's wedding wreath among the stars, where it still blazes from afar as the NORTHERN WREATH—a befitting tribute to beautiful Ariadne.[7]

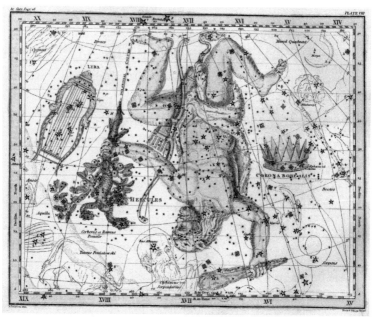

Jamieson Plate 8: Corona Borealis, Hercules, Lyra

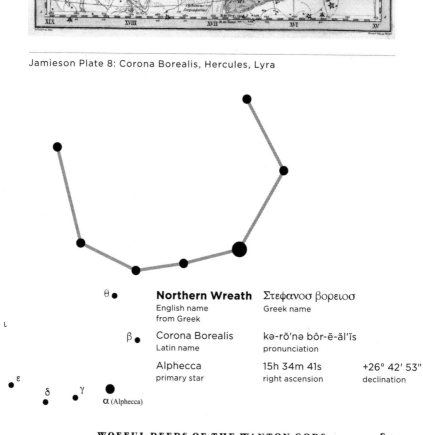

Northern Wreath
English name
from Greek

Στεφανοσ βορειοσ
Greek name

Corona Borealis
Latin name

kə-rō'nə bôr-ē-ăl'ĭs
pronunciation

Alphecca
primary star

15h 34m 41s
right ascension

+26° 42' 53"
declination

In the age of bronze, the belligerence of mortals and gods brought countless sorrows upon the Earth. But in the end, good prevailed. Zeus's seduction of Leda was shameful; but the resulting birth of the twins—Castor and Polydeuces—provided the world with two heroes who faithfully served both gods and men. Likewise, his abduction of Europa was a dastardly act, but it led in later years to the birth of a granddaughter—Ariadne. The sweet and gentle girl saved the lives of Theseus and thirteen other Athenian youths, while helping to rid the world of the monstrous Minotaur. As Dionysus' wife, she delighted the gods and gained their everlasting love.

Zeus—forever the philanderer—fathered other children by three of the seven sisters known as the Pleiades. But, once again, good prevailed in the end. The three children born of these trysts rose to greatness among immortals and mortals alike. Maia, the fairest of the Pleiades, brought forth the infant Hermes, who later flew through the sky as the messenger god so dearly loved by humans. Maia's sister, Taygeta, bore a son named Lacedaemon, who founded the kingdom of Sparta. Electra's son, Dardanus, became the father of the Trojan race. These children offered further proof that good may come from even the basest beginnings.

4

THE PRICE OF ARROGANCE

CONSTELLATIONS

ORION (Orion)
SCORPION (Scorpius)
CLAWS (Libra)
HARE (Lepus)
DOG (Canis Major)
HERALD OF THE DOG (Canis Minor)
RIVER (Eridanus)

STARS
DOG STAR (Sirius)
HERALD OF THE DOG STAR (Procyon)

The gods had come to disdain mortals because of their failure to offer proper respect and devotion. Arrogance, above all, provoked the deities, because it created strife among humans and made men believe they outshone all beings, mortal and divine.

ORION, THE HUNTER

One man, Orion, loomed large and mighty among mortals, but stumbled and fell into the same folly of conceit. In stature he stood like a giant, and his prowess as a hunter impressed all who saw him. Odysseus himself called Orion "the tallest, and far the most handsome" of men. "In his hands he held a club all of bronze" and the pelt of an animal—the trademark of a hunter.[1]

Orion came of age in the way of the warriors and hunters of Hellas. From childhood, they carry a simple array of weapons into the chase: a club, a pelt, and a knife. Although very young, they sharpen their senses by wandering far and wide, alone, through fields and forests. They learn to detect the slightest movements among the blades of grass or the leaves of trees. They listen for subtle sounds that betray the presence of prey.

As they grow strong and hardy, these hunters of Hellas become brave and bold. Some swagger as if they are brandishing weapons in battle. Their glowing confidence and sharp skills of observation lead them to excel in other areas as well. Their self-discipline and stamina pave the way for success in every endeavor.[2]

Hunters—young and old—often wander the winding woodland trails with one or more canine companions. The respect between dog and man is mutual, and their firm friendship is often portrayed in hunting scenes painted on pottery. Both play an important part in the chase and share in the feast that follows. Hunters highly

Orion depicted in the *Uranometria* of Johann Bayer, 1603. Courtesy of the Museum of History and Science Library, Oxford, England. Photo by author.

value their dogs, especially those of fine form and solid training. They depend on them, not just for stalking hares—the favored game—but for protection from large and lethal predators, such as wolves, boars, bears, and lions.

Like most Hellenes, Orion enjoyed hunting the hares that appear in abundance across the mainland and islands. He thrilled at the quick and lively chase that caused his heart to pound. He savored the succulent meat and valued the plush pelts that are sewn together for warm cloaks and bedding.

In the way of the woodsman, Orion carried the pelt of a large animal to ensnare the hare while his club came hurtling down. In his belt dangled a dagger for skinning and dressing the game.

After he roasted the meat on a spit, he used the knife to slice a portion to place on the burning coals as an offering to the gods. Only then did he stuff his ravenous mouth and toss a piece of meat beneath the pleading eyes of the hungry hound.

Rarely did Orion miss a chance to follow his dog in pursuit of prey. The canine would wander warily ahead—quick of eye and keen of scent, with paws toughened by years of roaming in rugged terrain. He was not solid in color, but spotted—a sure sign of good breeding. His hind legs stood long and strong, with plenty of spring for leaping, and perfectly formed for the sprint.[3]

On the trail, the hound trusted his sharpened sense of smell and followed his nose while sniffing the earth for a sign. For this, Orion favored a crisp morning in spring or fall—not in winter, when the frozen ground conceals the scent of prey; nor in summer, when heat disperses the smell from the sunbaked soil. Nor did the hunter delay until after dawn, when the scent of the previous night grows faint, and the rising wind of daytime whisks it away.

As they wandered the woods together, with the dog silently searching for the scent, Orion whispered a prayer for success—a pledge of first portions of meat for Apollo and Artemis, the twin gods of the hunt. Suddenly the dog trembled and shook in his tracks upon detecting a tempting smell. The hare, somewhere nearby, remained silent and still as a statue. To hide himself in a half-dug hollow or clump of grass, his head and body lay prostrate, with legs tucked underneath. His long ears lay flat on his back.

With a frenzied yelp, the dog shot along the line of scent, swerving to and fro like an un-fletched arrow. He followed close on the course while "barking freely" and running "fast and brilliant in the chase." At the last possible moment, the hare bolted and fled, with ears held high. Twisting and turning—teasing his tracker—he darted down shortcuts known only to him. He ran the poor dog through brambles and briars, then up—always up—steep slopes.

The cunning hare knew from experience to "cross brooks and double back, and slip into gullies or holes." More often than not, he outsmarted the hound. But Orion and the dog never seemed to mind. Tomorrow's dawn always brought another delightful chase.[4]

Nothing in heaven or earth could ever distract Orion from his love of the hunt, until the day he laid his eyes on the Pleiades. The darling girls—the seven daughters of Atlas and Pleione—had attracted many men, but none more deeply smitten than him. Like a woodsman after a wary deer, he pursued them with an obsession—but all in vain.

Before now, no maiden had escaped his affections. Some say even Artemis—the champion of chastity—almost fell for his manly charms. But the apprehensive Pleiades resisted his every impassioned ploy.

At last, the beleaguered sisters begged Zeus for protection. The god of sky and storm obliged, and changed them into a peaceful covey of pleasantly cooing doves. In later years, he granted them immortality among the stars, where they rest forever—nestled together in lovely company.[5]

Orion reeled at the rebuff. Deeply disappointed, he wandered back to the woods to follow his first love—the hunt. Now his pursuit of game became more excessive as he tried to bury his feelings for the Pleiades behind a show of bravado.

Loudly, Orion boasted that he could vanquish any beast—great or small. Gaea—Mother Earth—finally heard enough. She abhorred his blasphemous bellows against nature, and shuddered in fear that he might come near to destroying all the fauna she dearly loved.

So Gaea, who once sent the monstrous Typhon against the gods, summoned a Scorpion, colossal and cruel, to slay Orion. Like Aesop's cocky rooster, Orion crowed too loud and lured a predator upon himself.[6] The mighty hunter now became the prey.

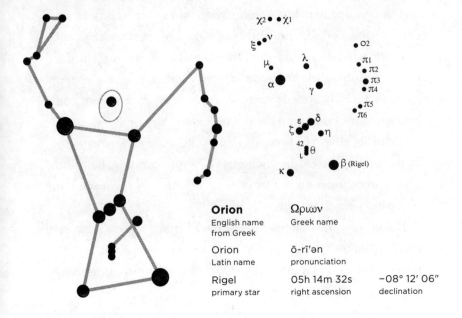

Orion
English name
from Greek

Ωρıων
Greek name

Orion
Latin name

ō-rī'ən
pronunciation

Rigel
primary star

05h 14m 32s
right ascension

−08° 12' 06"
declination

While Orion walked in the woods one day, the Scorpion—born of the earth—burst to the surface, showering dirt in every direction. On spying Orion, he bore down hard and fast, scurrying swiftly forward on eight arachnid legs, with pinchers snapping. Orion raised his thundering club and fended off the claws with lightning strokes. But soon the Scorpion brought his stinger into play.

The stinger stretched longer than Orion's club and dripped with poison more potent than the venom from a hundred vipers. Dodging the deadly dart, the embattled man struggled to reach the monster's head and crush it with a single savage blow of the club. As Orion pushed desperately forward, with eyes wide and watching in every direction but down, he tripped on a boulder and tumbled to the ground. In an instant, the Scorpion shuffled forth and pierced him through the heart. Thus, the mighty hunter fell to the "fiery sting" of a scorpion "proving mightier."[7]

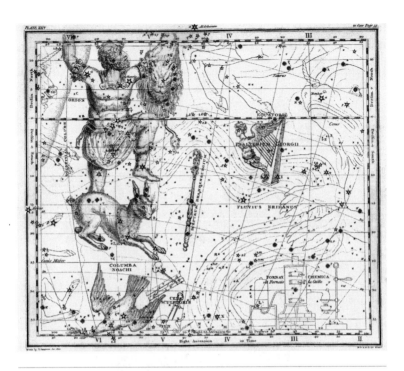

Jamieson Plate 24: Orion, Lepus

When ORION died, Artemis fell to her knees and grieved. She pleaded that the hunter might receive a home in heaven. Zeus complied. But because he despised immoderate pride in mortals, he placed the SCORPION in the cosmos as well. There, the beast forever follows Orion and offers a lesson for all who see—to avoid conceit for fear of a fall.

The massive Scorpion stretches so far across the celestial sphere that he occupies more than his fair share of the starry sky. To remedy this, his powerful pinchers became defined as a separate constellation, called the CLAWS.[8] As the Scorpion appears on the eastern horizon, Orion seeks shelter far to the west. But when the Scorpion is hidden from sight, Orion seems happy in his heavenly home, and no one shall ever "see other stars more fair."[9] With his

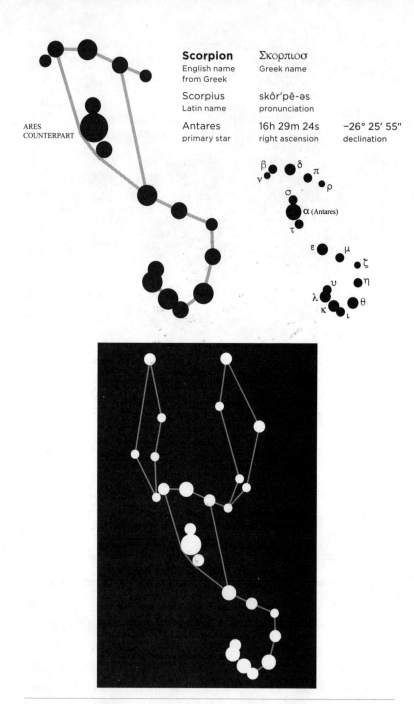

Scorpion
English name
from Greek

Σκορπιοσ
Greek name

Scorpius
Latin name

skôr′pē-əs
pronunciation

Antares
primary star

16h 29m 24s
right ascension

−26° 25′ 55″
declination

ARES
COUNTERPART

β
ν
δ
π
ρ
σ
α (Antares)
τ
ε
μ
ζ
η
υ
λ
κ
ι
θ

Scorpion and Claws

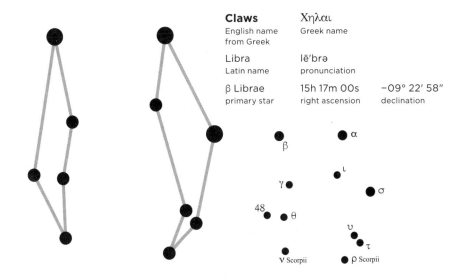

Claws
English name
from Greek

Χηλαι
Greek name

Libra
Latin name

lē′brə
pronunciation

β Librae
primary star

15h 17m 00s
right ascension

−09° 22′ 58″
declination

α

β

ι

γ

σ

48

θ

υ

ν Scorpii

τ

ρ Scorpii

enormous stature and radiant aura, as well as his position among the equatorial stars, he shines down upon all the people and places of the Earth.

From on high, Orion still pursues the Pleiades across the sky.[10] And yet, he prefers to hunt the celestial prey—with a pelt in the left hand, a club held high in the right, and a glittering knife tucked into his belt. Even the Bear "circles ever in its place and watches Orion" with suspicion.[11] But the hunter's attention is fixed on the HARE, who races ever underfoot, with ears erect.

Here, Hermes—the messenger god—set the Hare to honor his awesome agility and speed. Like his earthly counterparts, the Hare sits silent and still among the stars, ready to run when discovered. Then he darts away, managing to stay just beyond reach, as though he enjoys the hunt as much as does the hunter.[12]

Orion's spotted comrade—the DOG—also joins the chase. Bounding through the astral lights, he bursts forward on long legs toward his prey and stays close on the heels of the cunning Hare. Like Orion, the Dog's constellation burns brilliant in the night.

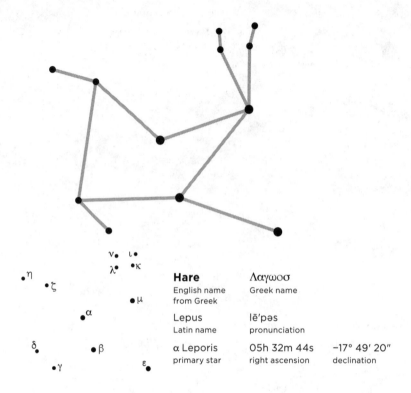

Hare
English name
from Greek

Λαγωοσ
Greek name

Lepus
Latin name

lē'pǝs
pronunciation

α Leporis
primary star

05h 32m 44s
right ascension

−17° 49' 20"
declination

On his muzzle he bears the brightest star in all the sky—the DOG STAR, that some call Sirius.[13]

Alongside the Dog frolics a playful pup—the smaller dog of Orion. The puppy's constellation is named HERALD OF THE DOG, because he rises above the horizon before the larger canine comes to view. The little dog likes to join the hunt, with his tiny tail happily wagging. But lacking the steady self-control of the older hound, he sometimes jumps ahead and spoils the stalk, or straggles behind and pauses to nip at flowers, as playful puppies will do. On his muzzle he carries a single star. On his flank glimmers a far more luminous light called HERALD OF THE DOG STAR, or Procyon. This star shines only slightly less bright than Sirius.[14]

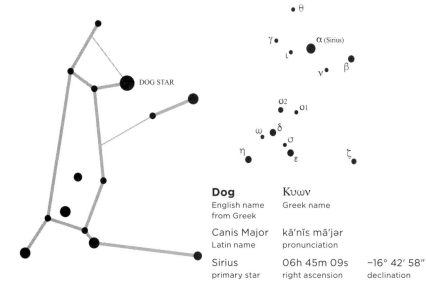

θ

γ •   α (Sirius)
   ι •
      ν •   β

DOG STAR

o2
   • o1

ω • δ
η   σ
   ε   ζ

Dog
English name
from Greek

Κυων
Greek name

Canis Major
Latin name

kā′nĭs mā′jər
pronunciation

Sirius
primary star

06h 45m 09s
right ascension

−16° 42′ 58″
declination

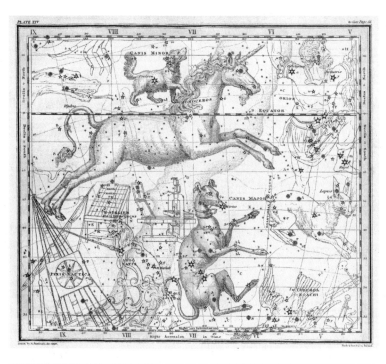

Jamieson Plate 25: Canis Major, Canis Minor

HERALD OF
THE DOG STAR

α (Procyon)

β

Herald of the Dog
English name
from Greek

Προκυων
Greek name

Canis Minor
Latin name

kā'nĭs mī'nər
pronunciation

Procyon
primary star

07h 39m 18s
right ascension

+05° 13' 30"
declination

THE FALL OF PHAETHON

High in heaven, Orion and his dogs hunt at the mouth of the "deep eddying" River Eridanus. The sad River—now a stream of stars— recalls the tragic fall of Phaethon, who, like Orion, was punished for his pride. Phaethon's father reigned on high as Phoebus Apollo (Bright Apollo)—the shining god of the sun.[15] But the boy was born a mortal child, raised by a caring mother—Clymene—and his older sisters, the lighthearted Heliades.

The sunny sisters loved to play among the trees on the river-bank, spreading their arms and fingers wide like the branches above or adorning their hair with the colorful autumn leaves. Sometimes they mimicked the cheerful songs of birds that flitted from twig to twig.

Phaethon joined their happy outings, and often paused with the girls to watch the wondrous sun as it rose and fell in the sky. First, it dispersed the darkness of early morning. Then it followed a golden arch over land and sea, across the broad, blue expanse. The children knew to be back home by the time it arrived on the western horizon.

From the beginning, Clymene told her children the truth: that the shining sun was indeed their father, Phoebus Apollo. She assured them that he smiled down on them daily and tanned their faces and gave their cheeks a rosy glow. Phaethon longed to be like his glorious father, but alas, he found himself forever bound to earth.

As the years passed and the restless boy grew bolder, he assumed the visage of invincible youth. With childhood fears left far behind, he decided to face his father at last. Straightaway he packed a bundle, slung it over his shoulder, and marched for many weary days to the base of Mount Olympus. Undismayed by the lofty peaks that jutted above, he began to climb, slowly and steadily, toward the top.

Upon surmounting the shining summit, he promptly found Apollo's resplendent palace. Without delay the brash boy pushed the gates wide open and entered. Soon he stood in the sun god's beaming presence.

Apollo welcomed Phaethon warmly, as a father to a son, and promised to hand him his heart's desire. The boy had no need to ponder his response. At once, he declared his undying wish, his lifelong desire: to drive the chariot of the sun for a day. From there he hoped to behold all the world on high, from earth to sea to sky.

Aghast at the reckless request, Apollo tried to dissuade him. None but himself, he said, could manage the fierce and fiery team of horses that draws the chariot across the cosmos. Not even Zeus would dare to take on the task.

The climb up to heaven was far too steep, Apollo warned, and the descent even more dizzying and dreadful. In addition, dangers dwelt among the stars. The Scorpion, the Crab, the Lion, and the Bull glowered at the deity's daily intrusion. How quickly they would maim a mortal youth—stinging, pinching, clawing, and goring!

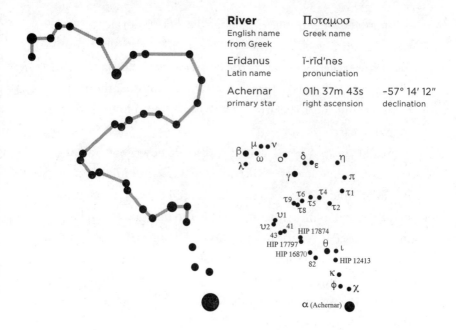

River
English name
from Greek

Ποταμοσ
Greek name

Eridanus
Latin name

ĭ-rĭd′nəs
pronunciation

Achernar
primary star

01h 37m 43s
right ascension

−57° 14′ 12″
declination

Phaethon only smiled at Apollo's words of warning, and the sun god saw for himself the folly of trying to withdraw his promise. Bowing his head and shaking it slowly, he whisked his son away to the sunlit stable in the east. There he prepared the blazing chariot and team of spirited steeds. Then he returned to his throne, sullen and troubled, to watch, with deep foreboding, the fateful day unfold.

Dauntless as ever, Phaethon leapt into the chariot and grabbed the reigns as if right at home. In an instant, the chariot jolted forward and shot into the air like an arrow, climbing at an alarming rate. Soon the team parted the puffy clouds and hurtled high into the blue sky. Fear gripped Phaethon now, and he ventured an anxious glance below. He watched aghast, in great distress, as homes, towns, islands, and mountains dimmed and finally disappeared in the distance.

As Phaethon began to shake with fear, the stormy horses

detected the driver's lack of control. Flushed in frenzy, they seized command and fought and jostled among themselves for a new path to follow. Without warning, they wheeled in every direction.

Into the highest heaven they flew, disrupting the peaceful abode of the stars. From there, they streaked toward earth in fiery descent and slammed across the mountaintops, leaving them leaping in flames. The lofty peaks—the sacred pinnacles of the gods—now burned and billowed in ruin. Even Olympus began to blaze. Forests followed—engulfed in violent infernos—and the waters of many rivers boiled away into suffocating steam.

Flora withered and died. Fauna scattered and fled; and humans wailed aloud as their homelands blackened in ashes. Gaea screamed for action from the immortal gods, demanding they save her lovely earth and all its living beings. Suddenly, Zeus stepped forward, furious at the widespread destruction and the chaotic disruption of his peaceful skies. He seethed, too, at the conceit of a brazen youth, who dared to assume the role of a god.

With a single stroke of his thunderbolt, he felled the tempestuous team of horses, and struck the rider and chariot from the sky. Phaethon, with "fire ravaging his ruddy hair," tumbled through the air with the burning wreckage—"as sometimes a star from the clear heavens . . . seems to fall." Into the River Eridanus he shot like a flash. With a deafening splash, he caused the waters to hiss and steam.

The naiads of the river sullenly retrieved his burned and broken body. Then they reverently buried the boy, "still smoking with the flames of that forked bolt." At last, in great solemnity the river nymphs engraved his epitaph: "Here Phaethon lies: In Phoebus' car he fared, and though he greatly failed, more greatly dared."[16]

Phaethon's inconsolable sisters, the Heliades, gathered close around him. There they mourned for their brother so steadfastly—without stirring from the spot—that they took root on the riverbank and assumed the form of poplars. With quaking leaves they shed sweet tears of amber that showered down upon the rippling waters and soft grass below. Zeus tried to assuage the sorrow of Phoebus Apollo by assigning the RIVER a place in the sky in memory of his children. Now a stream of stars stands in place of the once proud flow of water known as "Eridanus, river of many tears."[17]

BELLEROPHON, THE PROUD

Another young and haughty hero—Bellerophon—suffered a similar fate. None could deny that he had performed amazing deeds. First, he found and approached the wild and unruly winged horse, Pegasus. Soon after, he softened the stallion's rampaging heart with his gentle hand and soothing voice. Then he cautiously climbed on his back and rode the steed until they became like one in flight.

Now the hero and horse flew far and wide to find adventure. In the east, they engaged in battle with the she-warrior Amazons— who despised all men and even discarded their own infant sons. In the furious fight that ensued, Bellerophon vanquished many of the warlike women. Flushed with victory, he now pursued the deadliest demon of all. Leaping proudly on Pegasus' back, he flew aloft toward the menacing mountains to face the dreaded Chimaera.

The Chimaera—the odious daughter of Typhon—made her lair among the desolate heights and rocky crags. Only through the air could the horse and rider find her. Huge and hideous, her body was that of an agile mountain goat that allowed her to leap from peak to peak. She had the flesh-eating head of a lion and a tail that coiled and writhed like a snake, striking with venomous fangs. At night, the cruel beast incessantly stalked the valleys, "breathing out, terribly, the force of blazing fire."[18] She hated every living thing, but, most of all, abhorred the sight of humans.

As soon as she spied the horse and rider hovering above, she roared forth a wrathful flame that singed their hair and baked their skin. The two bellowed in pain but continued their pursuit. Circling at a distance, they dodged the fire and fangs while the young man pierced the creature's body with a rain of arrows. At last, she grew faint and plummeted to her fate in the valley below.

Now that Bellerophon had laid low the fierce Chimaera, he assumed himself invincible, with victory even over death. He began to imagine that he, who seemed immortal, must surely deserve a divine home. Climbing once more on Pegasus' back, he goaded the horse to soar through the sky, toward the summit of Mount Olympus. Here he expected to join an applauding pantheon of gods.[19]

As Bellerophon spurred his winged horse higher and higher above the clouds, Zeus delivered a woeful welcome—a gadfly—to render a painful bite on Pegasus' back. The horse began to buck wildly, dislodging Bellerophon and causing him to fall through the blue sky and down to the dusty earth. A thorn bush saved the tarnished hero from sudden death, but left him battered and bruised in both body and mind. Despised by the gods, and lacking his former strength and self-esteem, he wandered alone until his final days, "eating his heart out, and shunning the paths of men."[20]

In spite of all his amazing achievements, his fatal flaw—his arrogance—doomed him in the end. Once again, the deities delivered the stern warning, in no uncertain terms, that pride precedes a fall.

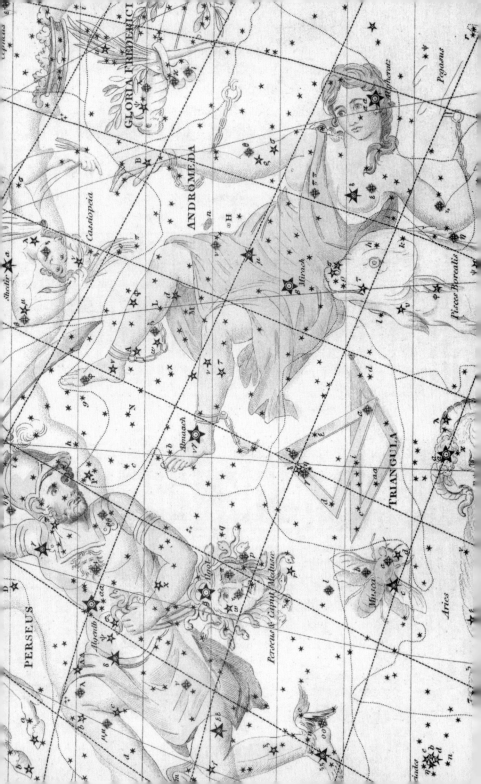

5

AN AGE OF HEROES

CONSTELLATIONS

Perseus (Perseus)

Andromeda (Andromeda)

Sea Monster (Cetus)

Cepheus (Cepheus)

Cassiopeia (Cassiopeia)

Horse (Pegasus)

Horse Head (Equuleus)

Dragon (Draco)

Kneeler (Hercules)

Hydra (Hydra)

Lion (Leo)

Crab (Cancer)

Arrow (Sagitta)

STAR

Gorgon (Algol)

Left: Jamieson Plate 3: Perseus, Andromeda

Orion stood tall with the stature and strength of a hero. Phaethon ventured forth with the confident glow of youth. Bellerophon vanquished dark forces in desperate battle. But the three fell victim to the same invisible foe: their own arrogance. They plummeted to their doom as punishment for pride, as a mighty oak crashes to the forest floor.

They were not alone. Mortal women succumbed to the same vices as men. You have heard of Medusa—a monster so ghastly that those who looked upon her froze into cold, lifeless stone. But the wretched creature was once a young lady of radiant beauty. Her lovely form, her flowing hair and enchanting eyes shone like a goddess and captured the gaze of men and women, of birds and beasts, of flowers and trees, and of immortal beings.

Even Poseidon—god of the sea—could not resist her charms. From the salty surf he daily splashed ashore and, at a distance, followed her footsteps. Her allure captured his heart and conquered his mind until he yearned for nothing more. Desperately, madly, he became entranced by the maiden Medusa.

But the admiration of deities and men destroyed her. Her sweet smile slowly sank and became a haughty sneer. At last, her pride proved so loathsome to the gods that Athena, who was also quite jealous, transformed Medusa into a gruesome ghoul. Her fair form now slithered on the ground like a serpent. Her flowing hair writhed as a tangle of poisonous snakes. Her enchanting eyes destroyed all who beheld her, so that she never again looked upon joyful life without watching it decay into dreadful death.

In her new form, hatred consumed Medusa, poisoning her heart and blackening her blood. Day and night, she shouted bitter blasphemies at the gods, but remained helpless to harm them. Instead, in hateful agony, she determined to destroy as many mortals as wandered within her dreary domain.

PERSEUS AND ANDROMEDA

At last, a trembling handful of heartbroken survivors came forward, forlorn, with heads held low in fear and sorrow. They had lost their loved ones—forever frozen in stone and condemned to stand as silent ornaments of anguish in Medusa's miserable lair. In despair, the survivors pleaded for the help of Perseus, a hero among them who shone with the unbounded boldness of youth. Together, they begged him to stalk and destroy the fiend. Perseus, the son of Zeus and Danae, stepped forth without further thought, as heroes will do. In haste, he pursued the dangerous quest and tracked Medusa to her haunts. Drawing near, he cautiously stalked past pillar after pillar of stone-cold mortals who had all shared the same misfortune.

To avoid a similar fate, Perseus wielded an assortment of weapons that had been given to him by the gods. A polished shield let him see Medusa's horrid reflection rather than face to face. Small wings attached to his feet launched him fast in flight, and a magic helmet rendered him invisible to her sight. Upon reaching the lair, he took to the air and searched with keen eyes until he spied her from above.

Then, like a storm, he burst upon Medusa with a lightning-fast sword. The monster swiftly spun toward the sound of the rushing wind. But before she could fix her eyes on him, Perseus, with a single slash, severed her scaly neck. The head thudded heavily to the ground, with glowering eyes still open and serpent hair writhing and snapping. Ever so carefully, he covered the lethal eyes and retrieved the head. In disgust, he dropped the loathsome trophy into a leather bag and made his winged departure through the air, far over the sparkling sea.

As the lifeblood flowed from Medusa's head, droplets from the blood-soaked bag fell and mingled with the seafoam far below. Poseidon mourned aloud and lamented the maiden's brutal punishment and cruel destruction. Unable to undo the dictates of fate, he decreed instead that Medusa's blood should join with his own seafoam to beget a beautiful being. For now that Perseus had released Medusa from her wretched existence, the god of the sea wished something good to come from the once-radiant woman he had loved. According to Poseidon's will, Pegasus—the winged horse, white as snow from muzzle to tail—was born of the blood and foam, and sprang from the salty sea.[1]

Meanwhile Perseus, rising in triumph on winged feet, soared for days and nights through the heavens, with the white-capped sea below and the approving gaze of the starry Pleiades above.[2] From his vantage point he beheld many strange and exotic kingdoms. In the far west he alighted in the land of the giant Atlas, whose wife, Hesperis, and daughters, the Hesperides, graced that tropical clime.

Perseus and the family fell into pleasant conversation. But alas, poor Atlas, more brawn than brain, spied the forbidden, blood-soaked bag. In an unguarded moment, with one curious peek at the hideous head and lethal eyes of Medusa, he froze into the North African mountain range that forever bears his name. With profuse apologies to Hesperis and her daughters, Perseus resumed his flight through the stars for three more days and nights. "Thrice did he see the cold Bears" in the northern sky, "and thrice the crab's spreading claws" among the constellations.[3]

On the fourth day, while following his course above the shoreline of Joppa, a tiny object, far below, captured his attention. Circling downward for closer inspection, he discovered a maiden, all alone, chained to the jagged rocks that jut from the surging sea. Her delicate form stood fully exposed to the salty air and rising tide. In rapid descent, Perseus rushed to her side.

The silent girl, cold and lifeless, seemed beautiful beyond perfection—the masterpiece, perhaps, of a sculptor inspired by the gods. But then "a light breeze ... stirred her locks and warm tears welled in her eyes."[4] The maiden, Andromeda, refreshed by the ocean air, shortly regained her senses. Startled at first by the stranger beside her, she soon set fear and shyness aside and told the young man her woeful tale.

Much like Medusa, her mother Cassiopeia had boasted of her own flawless beauty. Cassiopeia—too foolish to realize how foolish she was—even claimed to outshine the Nereids, the lovely sea nymphs cherished by Poseidon.[5] The god of the sea seethed at the insult, but Cassiopeia remained self-absorbed and ignored his vengeful wrath. She failed to consider that it was Poseidon who punished wayward sailors on the storm-tossed seas, and Poseidon who sent rumbling earthquakes and raging tsunamis to destroy coastal towns and entire fleets of ships.

The indiscretion of Cassiopeia now prompted Poseidon to unleash Cetus—an amphibious Sea Monster, prodigious and deadly. The voracious creature ravaged the land of Joppa, daily devouring unaware fishermen along with their boats and nets. He crunched the bodies and bones of shepherds and flocks as they roamed the rocky coastal paths. In gluttony and greed, he even snatched screaming guards from atop the city walls.

At last, the husband of Cassiopeia—the weak-hearted Cepheus, King of Joppa—saw no choice but to appease Poseidon with a sacrifice of royal blood to the monster Cetus. Soon Cepheus settled on a suitable victim: not himself—that would never do—nor his wife who had caused the whole misfortune; instead, he gave up their gentle, obedient daughter.

When the sobbing girl had finished her tale of impending doom, Perseus looked up, startled to see the barnacle-encrusted beast already "hurrying from the Atlantic water towards its

Andromeda Rocks, protruding from the Mediterranean waters at the ancient port of Joppa, mark the legendary site of Perseus' rescue of Andromeda. Jaffa, Israel. Photo by author.

maiden-feast."[6] The hero, smitten with love for Andromeda, promptly prepared for battle. In a flash, he leapt into the sky with winged feet and a gleaming sword in hand.

Suddenly, Cetus broke the surface of the sea. Perseus attacked at once. Time and again, from high and low, from front and back, from side to side and every angle he hacked and stabbed, gouging deep wounds while darting away from the gaping jaws and gnashing teeth. At last the loathsome leviathan began to sway, sluggish from fatigue and loss of blood, and finally collapsed with a deafening splash that nearly engulfed the anxious watchers crowded atop the city wall. The grisly deed now done, Perseus freed the fainting girl.

For many days and nights, the kingdom of Joppa reveled in victory with joyful shouts and jubilation, with festive processions and songs. Andromeda, shy and silent, received a warm welcome amid petals of flowers showered from the city wall. All the while, her mother pouted in the background, bewailing the mistreatment that left her ignored and unattended.

Although adored by her countrymen, Andromeda cared nothing for that, or for anything other than Perseus. The two vowed to leave Joppa behind, and, side by side, begin a new life in the seaside city of Argos. Cepheus and Cassiopeia, red in the face, railed against the thought of their daughter marrying the foreign youth. But Andromeda distrusted the judgment of those who had sacrificed her in order to save themselves. With her mind made up, she took Perseus' hand in hers.

Zeus, the proud father, saw much in PERSEUS that he admired and honored his heroic son in heaven. There, among the stars, Perseus is clad in a sturdy helmet and winged sandals, with a pol-

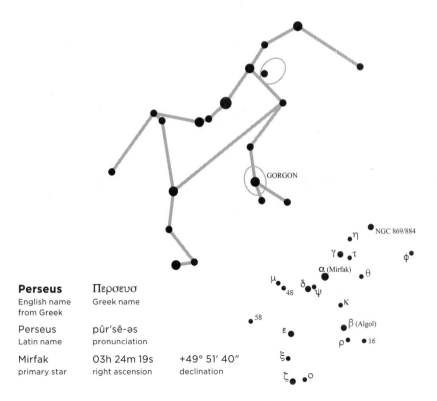

Perseus	Περσευσ
English name from Greek	Greek name
Perseus	pûr′sē-əs
Latin name	pronunciation
Mirfak	03h 24m 19s	+49° 51′ 40″
primary star	right ascension	declination

ished shield slung on his back. In his right hand he raises a flaming sword. In the left he carries the ghastly, ghoulish head of Medusa, marked by a beam of light called GORGON. Behind the head trail two stars of writhing serpent hair.[7]

To complete the scene, lovely ANDROMEDA swoons in her chains near the hero's upraised sword. The SEA MONSTER, with gaping jaws, rises menacingly from the stars beneath her feet, rushing upward, propelled by massive fins. CEPHEUS and CAS-SIOPEIA are there as well, forever watching the dread event from a cowardly distance.

Cepheus wears a felt cap that points forward, in the Asian fashion of his country.[8] He recoils in horror at the sight of the approaching fiend. Cassiopeia, saucy and scantily clad, sits on the edge of her throne with arms upraised in suspense. As punishment to the haughty queen, Zeus ordained that her constellation

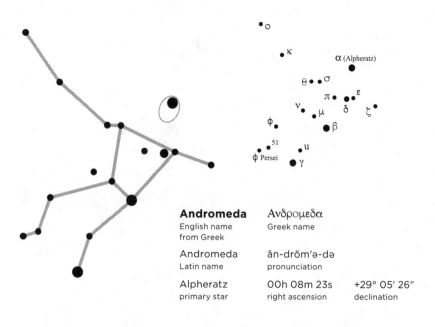

Andromeda
English name
from Greek

Ανδρομεδα
Greek name

Andromeda
Latin name

ăn-drŏm'ə-də
pronunciation

Alpheratz
primary star

00h 08m 23s
right ascension

+29° 05′ 26″
declination

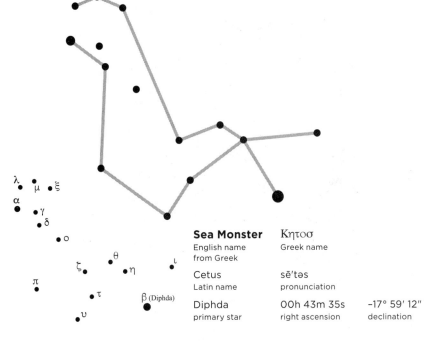

Sea Monster
English name
from Greek

Κητοσ
Greek name

Cetus
Latin name

sē′təs
pronunciation

Diphda
primary star

00h 43m 35s
right ascension

–17° 59′ 12″
declination

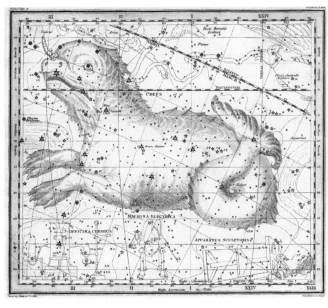

Jamieson Plate 23: Cetus

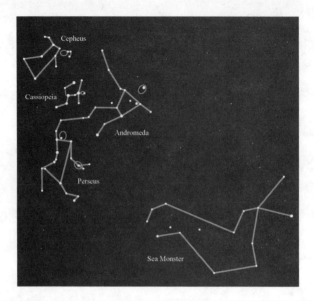

Andromeda Scene, showing the relative positions of the constellations in the sky.

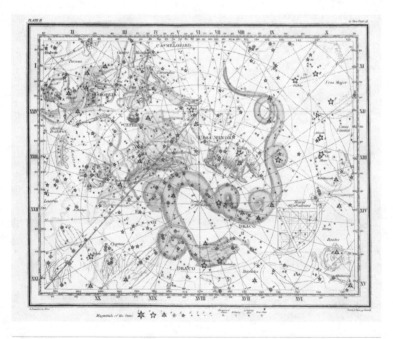

Jamieson Plate 2: Cepheus, Cassiopeia, Draco, Ursa Minor

ANCIENT SKIES

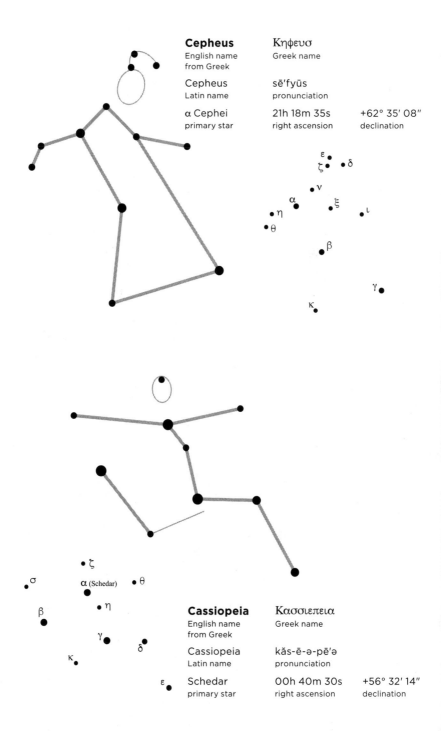

Cepheus
English name
from Greek

Κηφευσ
Greek name

Cepheus
Latin name

sē'fyūs
pronunciation

α Cephei
primary star

21h 18m 35s
right ascension

+62° 35' 08"
declination

Cassiopeia
English name
from Greek

Κασσιεπεια
Greek name

Cassiopeia
Latin name

kăs-ē-ə-pē'ə
pronunciation

Schedar
primary star

00h 40m 30s
right ascension

+56° 32' 14"
declination

should turn head-downward as she travels across the sky. In this awkward position, Cassiopeia can be neither proud nor modest. For "she headlong plunges like a diver" into the sea, as her stars set in the west.[9]

PEGASUS, THE WINGED HORSE

As for Pegasus, the winged horse that sprang from the sea became famous far and wide for his beauty and bravery. He often hurried to the aid of heroes involved in desperate quests. He carried Bellerophon aloft to vanquish the vicious Chimaera. Then, after Zeus dislodged the presuming youth from his back, Pegasus continued his ascent "and came to the immortals" on Mount Olympus.[10] On the shining summit that overlooks the world, he joined the livery

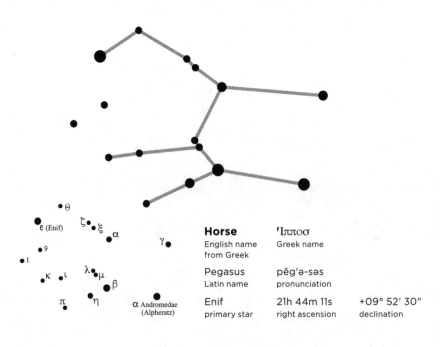

Horse	Ἵπποσ
English name from Greek	Greek name
Pegasus	pĕg'ə-səs
Latin name	pronunciation
Enif	21h 44m 11s +09° 52' 30"
primary star	right ascension declination

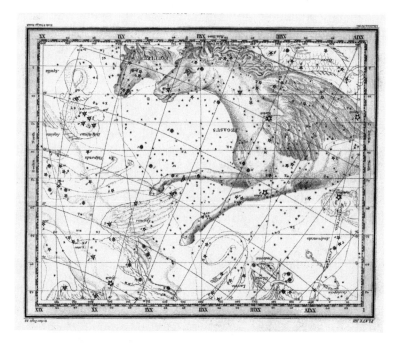

Jamieson Plate 12: Pegasus, Equuleus

of Zeus. Soon he served as the god's most trusted steed, bearing his flashing thunderbolts throughout the starry heavens.

Years later, when the mortal horse died, Zeus provided Pegasus with stars of his own. His constellation, simply called the HORSE, gallops swiftly, with tremendous spirit, through the sky.[11] By his side, within easy reach of his muzzle, runs his favorite foal, named Celeris.

During his earthly life, the sleek colt Celeris stood second only to Pegasus in splendor and speed. Celeris also displayed the wondrous traits of his distinguished maternal grandfather, Chiron the centaur.[12] Among the stars, only the head of Celeris is visible as he joins his sire in the celestial race. For this reason, his constellation is called the HORSE HEAD.

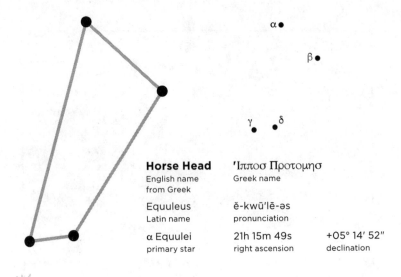

Horse Head
English name
from Greek

Ἱπποσ Προτομησ
Greek name

Equuleus
Latin name

ĕ-kwū′lē-əs
pronunciation

α Equulei
primary star

21h 15m 49s
right ascension

+05° 14′ 52″
declination

THE LABORS OF HERACLES

In Argos, the love between Perseus and Andromeda grew profoundly with the passing years, and they brought many children into the world. Their oldest son, named Perses, became the eponymous ancestor of the Persians. Other sons received renown as forebears of the Mycenaean kings. One of these—Electryon—reigned justly as ruler of Argos. Electryon's daughter, Princess Alcmene, became Zeus's lover and bore him a son.

After the birth of the baby, Alcmene soon saw that her little boy—Heracles—was no ordinary child.[13] He quickly gained the strength of a grown man, and showed stunning signs of his divine lineage by strangling poisonous serpents and performing other powerful deeds. Upon coming of age, Heracles followed in the heroic footsteps of his great-grandparents, Perseus and Andromeda.

Of all his amazing exploits, Heracles attained greatest fame for accomplishing the Twelve Labors, a series of seemingly impossible feats. The first labor required him to slay the Nemean Lion—a man-killer with a hide no weapon could pierce. The frightened villagers of Nemea had suffered dreadfully, with countless lives lost to the bloodthirsty beast. The brutal lion relished human flesh, but found far greater pleasure in the chase and the kill.

Many valiant warriors had tracked him down only to fall prey to his ripping claws and rending teeth. But Heracles remained undaunted. He had defeated a dozen lions with the help of his heavy club and arrows of bronze. With little trouble and little fear, he found the lair, littered with bones, and emitting a foul and suffocating stench. The lion had no reason to hide his presence. He always welcomed another woeful victim.

As soon as Heracles entered the dreary den, he discerned the huge, shaggy hulk in the darkness and launched a volley of arrows in rapid succession. In dismay, he watched as each bounced off the impenetrable hide and clattered to the cold cavern floor. Furious at the brazen assault, the lion lunged at Heracles with claws poised and fangs bared. As the two interlocked in deadly embrace, the lion slashed and snapped at his assailant time and again. But the powerful man held tight to the mane and slipped his muscular arms around the enormous neck, slowly choking the lion's life away.

With victory at last, Heracles sat on the ground with a heavy thud and tended to his open wounds. After catching his breath, he skinned the lion with its own sharp claws—the only tool that could sever the heavy hide. From that day on, the lion-skin cloak became the hero's most treasured possession and trademark.

Scarcely had the people of Nemea received the joyful news before Heracles was summoned to an equally trying task. The murky marsh of Lerna, that lay near Argos, had long been haunted by Hydra—the nine-headed swamp serpent, vile and vicious. The monster was one of the ferocious offspring of the "terrible, outrageous, lawless" Typhon—who had once attempted to devour the gods. From deep within her loathsome lair, the Lernaean Hydra slithered in ceaseless search of unwary wanderers.

For most of a day, Heracles struggled through the swamp in stagnant water up to his neck. He held a pine-torch high above his head to light a path through the forbidding fog and gloom. Suddenly he detected a slight ripple on the water. As he inched slowly forward, Hydra burst to the surface behind him. Heracles turned at once and swung his sword, severing one of her snapping heads.

In horror, he watched as two more heads grew promptly in its place. Quickly he lopped off another, but two more emerged. In desperation, he dispatched a third and thrust his torch onto the bleeding wound. The burning neck sizzled and boiled with a terrible stench, then fell with a splash, cauterized and lifeless. Hydra all the while struck with venomous fangs from every direction.

Hercules battling Hydra. Vase painting, c. 525 BC. Courtesy of the John Paul Getty Museum, Getty Villa, Malibu, California. Photo by author.

But one at a time, Heracles managed to dismember and singe each neck.

When only three heads remained, Hydra's companion, the Crab, sidled through the mire unseen. With piercing pincers he tore a nasty gash in the hero's foot. Heracles shouted in pain as he brought the wounded limb crashing down to shatter the shell of the Crab. He then turned his full fury on Hydra, and soon reduced the foe to one immortal head, which could not be killed. This he dispatched and concealed under a massive boulder, where it remains to this day.[14]

As news of Heracles' prowess spread through the land, he received frantic calls to more dangerous missions. Among these was his fight with the foul, flesh-eating Stymphalian birds. The menacing, man-eating flock had descended on the people of the Peloponnesus like a dark and ominous cloud.

Day after day, the grim, ravenous birds gorged on men, women, and children, until none dared to venture outside their homes. Whole families began to die of starvation as they hid in terror under beds or blankets. Heracles rushed to the rescue. To aid him in the quest, he chose his strongest, straightest Arrow, dipped in the poisonous blood of Hydra.

Draping his lion skin over his shoulder to serve as a shield, he drew his bow and slayed first one bird, and then another, retrieving his Arrow each time. By the end of the day, he had defeated most of the horrid fowl and chased the remainder far away. Now the starving people of Stymphalia, gaunt but grateful, emerged from cover. With hoarse and feeble voices, they cheered their deliverer.

Heracles had little time to tarry, but pushed ahead to his final labors. In the land of the Hesperides there dwelt a Dragon named Draco—a gargantuan servant of the goddess Hera. Hera—the wife of Zeus—heartily despised Heracles because Zeus had sired him

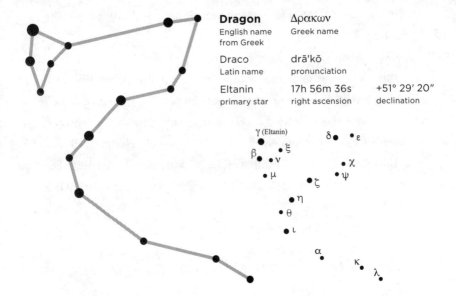

Dragon
English name
from Greek

Draco
Latin name

Eltanin
primary star

Δρακων
Greek name

drā'kō
pronunciation

17h 56m 36s
right ascension

+51° 29' 20"
declination

γ (Eltanin)

β
ν

μ

δ
ε

ξ

χ
ψ

ζ

η

θ

ι

α

κ
λ

with another woman. So Hera incessantly searched for ways to harm him, and Heracles responded in kind.

If Heracles could subdue her serpent, Draco, then he could help himself to the golden apples of the Hesperides—Hera's prized possessions. Hera had received the apple tree as a wedding gift when she married Zeus and proudly planted it in her garden on the African shore. Here it flourished under the constant care of Atlas' charming daughters—the Hesperides.[15] But the young maidens could not resist nibbling the forbidden fruit. Hera, therefore, employed the Dragon—Draco—as a more reliable watchman.

Draco, with his overpowering size and blazing eyes, proved to be one of Heracles' greatest opponents because he remained ever vigilant—never sleeping.[16] Still, the fearless hero bolted into the fray, without hesitation, and wrestled the scaly serpent to the ground. With immense power, Draco writhed and attempted to wrap his attacker in coils and crush his life away.

All that day the two struggled to gain a commanding grip, until, at last, Heracles managed to pin the Dragon's slippery head to the ground. Immediately he brought his heavy club down to shatter Draco's spine. As the lifeless serpent continued to wriggle and squirm, Heracles seized a branch from the apple tree—loaded with golden fruit—and made his departure.

Zeus applauded the astounding success of his stormy son. And when Heracles came to the end of his earthly days, the god placed him high in heaven with his other son, Perseus. In memory of one of Heracles' many battles, he appears clad in his lion-skin cloak, kneeling in desperate struggle with Draco the DRAGON. His left foot pins the serpent's head while his right knee rests on the scaly back. His club hovers high, clutched in the right hand, ready to deliver the lethal blow.

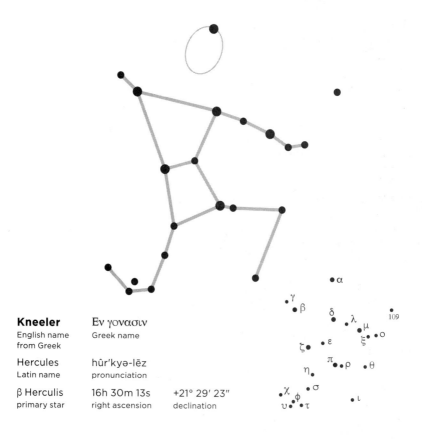

Kneeler	Εν γοναϲιν	
English name from Greek	Greek name	
Hercules	hûr′kyə-lēz	
Latin name	pronunciation	
β Herculis	16h 30m 13s	+21° 29′ 23″
primary star	right ascension	declination

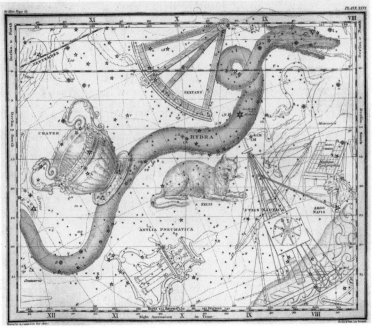

Jamieson Plate 26: Hydra, Crater

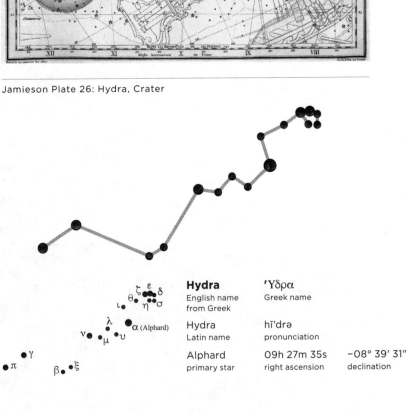

Hydra
English name
from Greek

Hydra
Latin name

Alphard
primary star

ʹΥδρα
Greek name

hī'drə
pronunciation

09h 27m 35s
right ascension

−08° 39′ 31″
declination

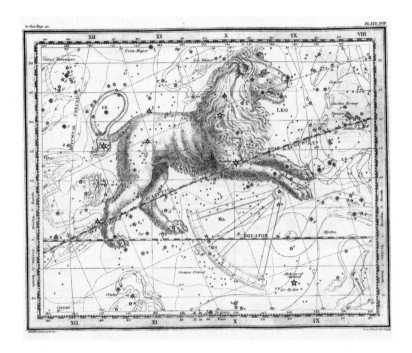

Jamieson Plate 17: Leo

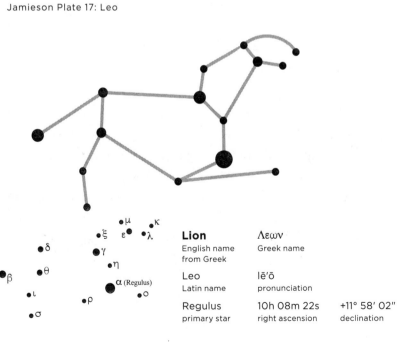

Lion
English name
from Greek

Λεων
Greek name

Leo
Latin name

lē'ō
pronunciation

Regulus
primary star

10h 08m 22s
right ascension

+11° 58' 02"
declination

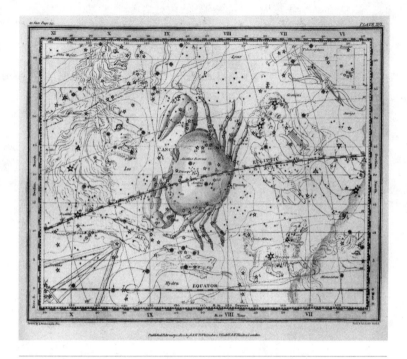

Jamieson Plate 16: Cancer

Because of the position in which he is poised, the constellation of Heracles is called the KNEELER. Lurking farther beneath him is HYDRA, slithering through the stars with her one immortal head still intact. Heracles' other enemies—the crouching LION and the stealthy CRAB—also appear nearby in the nighttime sky, as does his trusty ARROW that slew the Stymphalian birds.

The honor that Zeus bestowed on Heracles, Perseus, and Andromeda showed that the gods are as quick to favor those who serve them and their fellow man as they are to punish the proud. These heroes' lives reflected courageous service without conceit. For this, they remain the most famous of mortals.

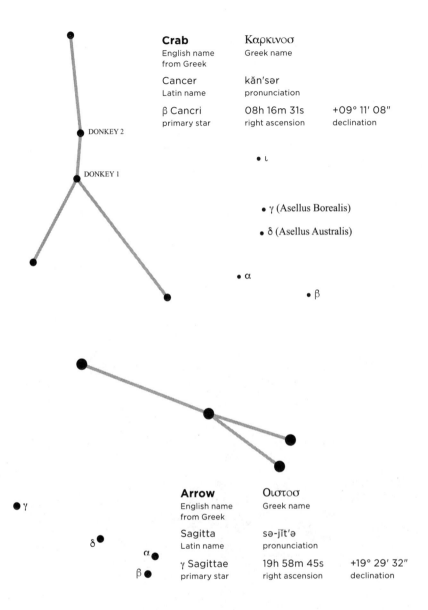

Crab
English name
from Greek

Καρκινοσ
Greek name

Cancer
Latin name

kăn'sər
pronunciation

β Cancri
primary star

08h 16m 31s
right ascension

+09° 11' 08"
declination

DONKEY 2

DONKEY 1

• ι

• γ (Asellus Borealis)

• δ (Asellus Australis)

• α

• β

• γ

δ •

α •

β •

Arrow
English name
from Greek

Οιστοσ
Greek name

Sagitta
Latin name

sə-jĭt'ə
pronunciation

γ Sagittae
primary star

19h 58m 45s
right ascension

+19° 29' 32"
declination

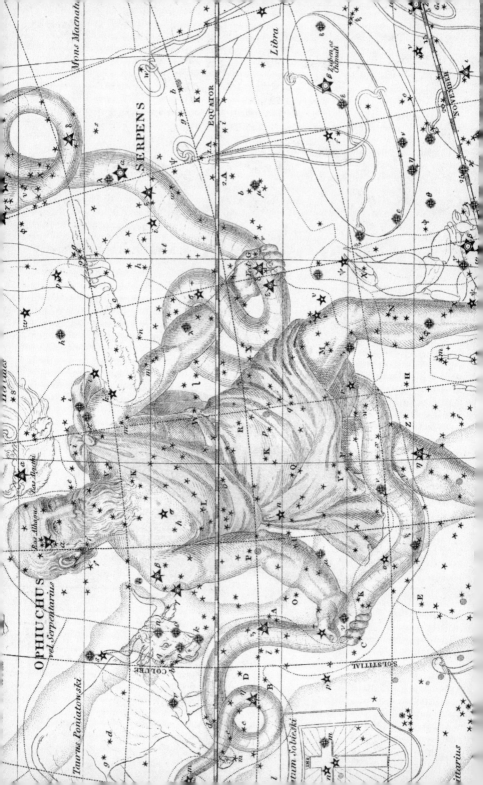

6

THE NOBLE ARGONAUTS

CONSTELLATIONS
—
RAM (Aries)
ARGO (Argo Navis)
LYRE (Lyra)
SNAKE HOLDER (Ophiuchus)
SNAKE (Serpens)
TWINS (Gemini)

Left: Jamieson Plate 9: Ophiuchus, Serpens

THE ARGONAUTS AND THE QUEST FOR THE GOLDEN FLEECE

Heracles is famous for the Twelve Labors, but he also searched for other far-flung adventures. Among them, he joined Jason's crew of Argonauts to sail the Aegean and Black seas and to capture the Golden Fleece.[1]

Other heroes also committed themselves to the quest. The crew included Asclepius, the ship's physician; Orpheus, the famed musician; Peleus, the father of Achilles; Laertes, the father of Odysseus; the twin brothers, Castor and Polydeuces; Argus, the builder of the ship *Argo*; and Tiphys, the talented helmsman.

Tiphys took on the critical task of steering the ship, because he was "expert in predicting rising waves on the broad sea; expert too in predicting storm winds, and in determining a course by sun or star." His lookout, Lynceus—whose name meant "lynx-like"—aided in navigation. Lynceus had the eyes of a lynx—the keenest vision of the crew.[2] Day and night he searched the sea and sky for telltale signs, with eyes squinted against gusts of wind and glare upon the water.

The crew embodied an excellent assembly of men of *arete*. Wherever they wandered, admiring throngs "marveled as they beheld the beauty and stature of the preeminent heroes."[3] Beyond their physical form and noble bearing, they loomed as men of learning. Some had gained the gift of knowledge from the wise centaur, Chiron.

Like many enlightened mortals, they excelled in music as well. On more than one occasion, they "sang a hymn to the accompaniment of Orpheus' lyre in beautiful harmony, and round about them the windless shore was charmed by their singing."[4] Above all, they stood tall as spiritual men who remained ever reverent and devoted to the deities.

The journey of the Argonauts arose from events that unfolded many years earlier in the kingdom of Thebes. The Theban queen had given birth to twins—a son and daughter—named Phrixus and Helle. But the king abruptly sent his wife away at the insistence of his mistress, who soon wore the queen's crown.

The new queen despised her stepchildren and conspired to see them put to death so her own son would one day inherit the throne. To this end, she demanded that her husband sacrifice his own children. This, she claimed, would appease the deities that had sent a famine to plague the land. Day and night, the wicked queen screamed at the king to slay Phrixus and Helle and save the kingdom's starving subjects.[5]

At last, the king caved in to his wife's constant carping, and commanded his royal servants to place the trembling children upon the pyre. All the while, the evil queen, unseen, sneered in the shadows. As the servants lit the loathsome fire, a splendid Ram with a Golden Fleece, sent by Hermes, suddenly darted through the chamber door.

Lowering his head, the Ram lifted the children onto his back and fled toward a safe haven in the east. Fast and far they flew. But as the Ram swam the swirling waters that divide Europe from Asia, poor little Helle, frail and fainting, slipped into the gloomy depths and drowned. On her behalf, the inhabitants along the coast later called the channel Hellespont.

Phrixus grieved deeply for his twin sister and cried a shower of tears while clinging tightly to the Ram's woolly back. After a weary, woeful journey, the two arrived on the eastern shore of the Black Sea, at the port of Aea in the kingdom of Colchis. The exhausted Ram, his mission fulfilled, submitted himself to Phrixus for a sacrifice to the gods—his last selfless act.[6]

In obedience, the boy, with sorrow heaped on sorrow, made the somber offering. Then he placed the Ram's Golden Fleece in a sacred grove to serve as a shrine. There it hung in a "huge oak tree . . . like a cloud that glows red from the fiery beams of the rising sun."[7]

Phrixus lived his remaining years in far-off Colchis, while Thebes and all of Hellas withered in drought—by divine decree—for their horrid deed and the death of Helle. Meanwhile, Phryxus' younger kinsman—Jason—was born to the king and queen of Thessaly and mistreated in a similar manner. When Jason was only an infant, his uncle usurped his father's throne and plotted to kill the child to secure his own spurious claim. So Jason was rushed to a cave for safekeeping and raised by the kindly centaur, Chiron.

When Jason reached adulthood, he returned to confront his uncle, the king. But the cunning uncle averted the young man's wrath and lured him into a dangerous quest that was sure to seal his doom. Jason brimmed with youthful boldness and could not say no to the venture.

The king and most of his subjects believed that the deadly drought would end, and the land would flourish again, if a champion only retrieved the Ram's Golden Fleece from Colchis. Thus the cruel and crafty uncle silently schemed: If Jason somehow succeeded in returning the Fleece, then the kingdom would thrive. But if he died in the attempt, the king would at least be rid of a threat to the throne.

Jason surmised the king's motive but remained undaunted, delighted by the chance for adventure and pleased to take on the task. Like him, his friends—the heroes of his generation—jumped at the chance to join him.

Soon, Argus began to build a seaworthy ship for the distant journey. First he crafted a keel of heavy oak and laid it on the pebble beach. To this he fastened beams and braces of hardwood, then heavy planking to hold back the furious slap of thundering waves.

Shipbuilding tools typical of the ancient Mediterranean: double-bladed axe, adze, linen chalk line, chalk, auger, and two types of mallets. Marshall Collection, Loma Paloma, Texas. Photo by author.

He carved a stern that arched above from behind, and a stout bow that thrusted forth from the foredeck. He chose a mast that was sturdy, straight and true, and planed it to perfection. He stepped the mast into the hull and wedged it with a heavy hammer, then straightened it with stays—side to side, and fore and aft.

Now he fitted a yardarm to the mast, with blocks and halyards. He shaped a sail and sewed it—triple-stitched—so it would billow well and hold in a heavy gale. He furled the sail on the yardarm to await the launch, while he fashioned a sturdy pair of steering oars and fifty more for rowing. He chose two stones for anchors and trailed them from the stern with two tightly braided ropes. With these the ship, named *Argo*, could rest stern-to-beach and not drift off with the ebb and flow of the tide.[8]

As Argus assembled the ship, Jason trained the crew and assigned specific tasks. At sea, Tiphys would steer from the stern while Lynceus watched from the bow. The others would work the sail when days were fine and the wind blew from behind. At other times, all hands would man the oars and row ahead toward some distant shore.

On making landfall late in the day on a foreign coast, the men would secure the ship and scout for danger. Once assured of safety, some would scoop up leaves for bedding, while others picked up piles of wood. Two would take up the task of sparking a fire by "twirling sticks" between their palms.[9] These men also served as cooks—a most important post. Their success or failure meant a healthy, happy crew or one that grumbled with stomachs growling.

With the ship and restless men ready to go, and the sea rising to high tide, the time came at last to launch the *Argo*. Under blue skies and fair winds, the crew winched and shoved the heavy wooden hulk across the pebble beach and into the swelling surf. Then they steadied the ship, "drew the sail to the top of the mast, and let it down from there. A whistling breeze fell upon it, and on the deck they wound the lines separately around the polished cleats and calmly sped past the long Tisaean headland. . . . A steady wind bore it ever onward."[10]

When the wind slackened that afternoon, the men bent to the oars. "Rowing proceeded tirelessly" as they plied the frothy waters of the Aegean Sea. Swiftly, they left Hellas behind in their wake as a hazy sliver on the western horizon.[11]

The trip began well, but trouble followed. One day, while seeking supplies on the island of Lemnos, Heracles' faithful servant failed to return to camp. Frantically Heracles searched for the youth along the shore and deep into the forest. Through the night and well past dawn he persisted, but in vain:

Soon the morning star rose above the highest peaks, and the breezes swept down. And at once Tiphys urged them to board and take advantage of the wind. In their eagerness they boarded right away, drew the ship's anchors up on deck, and pulled back on the halyards. The sail bulged in the middle from the wind, and far out from the shore they joyfully were being borne.[12]

Only after the island was far behind them did the men discern the absence of Heracles. Above and below the deck they searched, but found no trace. Downtrodden, they saw no option but to keep the course determined by the wind and continue without him.

As the crew probed into foreign waters and ran ashore on strange islands, they encountered unexpected dangers. Often they battled with hostile bands that resented their bold intrusion. Then, when they entered the tempestuous strait of the Hellespont, and the turbulent waters of the Sea of Marmara, the weather turned against them. The men grew weak and weary at the oars as they worked their way through lashing waves and heavy swells.

At last they beached the boat, staggered ashore, and collapsed in the sand. For twelve days and nights they rested, and recouped their strength while waiting for the storm to subside. On the eleventh evening, a halcyon shore bird hovered above the drenched and dejected men as they sat in the wet sand near the *Argo*. The halcyon arrived as a favorable sign, sent by Rhea, to herald fair weather for sailing. For this, the men offered grateful praise.[13]

The titan Rhea is the daughter of Gaea—Mother Earth. As such, Rhea holds some control over earth, wind, and sea. She is also the mother of Zeus, Poseidon, and other powerful gods. To be in her good graces is no small matter, but she must be appeased. Most people believe she lives among the Phrygians when she walks upon the Earth. As fate would have it, the Argonauts had happened to find safe haven from the storm on the Phrygian coast, in the shadow of Mount Dindymum.

Before resuming the journey, they carved a wooden likeness of Rhea and carried it up the mountain slope on their shoulders. They placed it in a sacred grove of ancient oaks and stacked flat stones for an altar. Donning wreaths of oak leaves, they offered a savory sacrifice.

As Jason continued to pray and pour libations, Orpheus taught the men to perform the leaping dance in armor—beating their swords against their shields in the manner of Phrygians who worship the matron goddess. Rhea's closest companions, the Corybantes, praise her in a similar way as they shout and twirl the whirring rhombus and beat the resounding tympanum.[14]

When the Argonauts received further signs that Rhea had blessed their voyage, the men's spirits soared. Now they gladly returned to the oars and worked their way through the Sea of Marmara and the narrow passage of the Bosporus.[15] After many more days, the arduous outbound journey ended at the distant port of Aea in Colchis, on the farthest reach of the Black Sea.

But there they faced fierce opposition. Aeetes, the king of Colchis, raged within when he heard that Jason intended to take the Golden Fleece to Hellas. The Fleece epitomized the glory and pride of his kingdom. It hung high, in the holiest place, and its value surpassed that of his total treasury.

Still, Aeetes quickly adjudged the tough and determined Hellenic crew that had come so far to tower before him. He feared the result of an outright rejection. So, instead, he freely offered the Fleece to Jason, if the youth would only succeed at a simple test of strength and skill. All he had to do was lash a heavy oak-tree yoke atop the bulging necks of two enormous fire-breathing bulls, then force the team to plow a furrow. Next he needed to plant the teeth of a dragon, and slay the angry army of men that would sprout forth from the dreadful seeds.

Rather than wagging his head and slinking away, as most men would do, Jason shrugged his shoulders, accepted the challenge, and pondered the best way to bring it about. From a balcony above, the king's sorceress daughter—Medea—witnessed the wager while her longing eyes followed Jason's every move. To her, Jason looked like Sirius—the shining star, the most luminous of all—"which rises beautiful and bright to behold."[16] Soon, Medea, the enchantress, became the enchanted.

Throughout the night, the anxious crew remained awake and watched the constellations of the Bear and mighty Orion. As they crossed the sky, they marked the passage of time until the dreaded contest at dawn. Meanwhile, Medea came to Jason in darkest night to reveal her love and to offer help with the deadly tasks.[17] First, she handed him an ointment that would fend off the fiery breath of the bulls. Then she described how he should defeat the dragon-tooth warriors by simply throwing a stone among them.

Enthralled by the love of the softhearted sorceress, and encouraged by the knowledge he now possessed, Jason prepared for the contest with confidence. At dawn, when the laughing king unleashed the bulls, Jason stepped forward, impervious to the flames that roared from their mouths. Promptly, he yoked them and forced them to plow a furrow while he sowed the seeds of dragon teeth. While he waited and watched for the fearsome army to rise from the furrows—like so many stalks of grain—he swelled himself with courage in the way of the warrior: "He flexed his knees to make them nimble, and filled his great heart with prowess, raging like a boar."[18]

Firmly he planted his feet and faced the "earthborn men" as they sprang forth to battle. As Medea had instructed, Jason hefted a heavy rock and hurled it among them. At once, they fell into confusion and turned on each other in furious slaughter. Jason now bolted forward, "as when a fiery star springs forth from heaven bearing a trail of light." In a flash, he vanquished the remaining foes.[19]

Aeetes shuddered, aghast at Jason's baffling success. But he was far from ready to concede defeat. Instead, the deceitful king delayed and offered to hand him the Fleece the following day. Then, secretly, he rallied his royal army and prepared to overpower the band of Hellenes.

Again Medea intervened. That night, she used a spoken charm and a potion of her own concoction to drug the great snake that guarded the sacred grove.[20] As the serpent fell into peaceful sleep, she showed the endangered crew the way to escape, unseen, with the Golden Fleece.

With Aeetes' army hot on their heels, the Argonauts flew to the ship and fled by sea. They brought Medea aboard as well, to protect her from her wrathful father. Heading swiftly westward, they

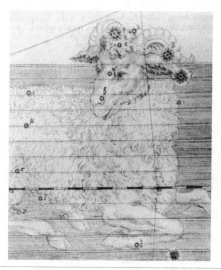

Aries depicted in the *Uranometria* of Johann Bayer, 1603. Courtesy of the Museum of History and Science Library, Oxford, England. Photo by author.

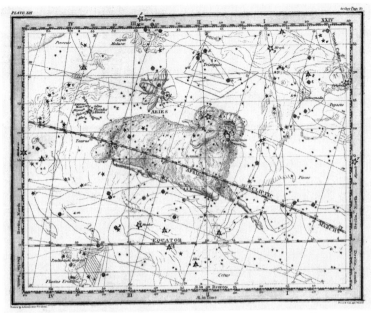

Jamieson Plate 13: Aries

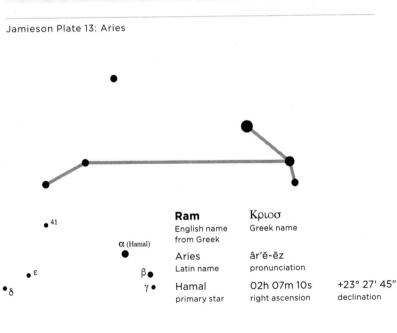

Ram
English name
from Greek

Κριοσ
Greek name

Aries
Latin name

âr'ē-ēz
pronunciation

Hamal
primary star

02h 07m 10s
right ascension

+23° 27' 45"
declination

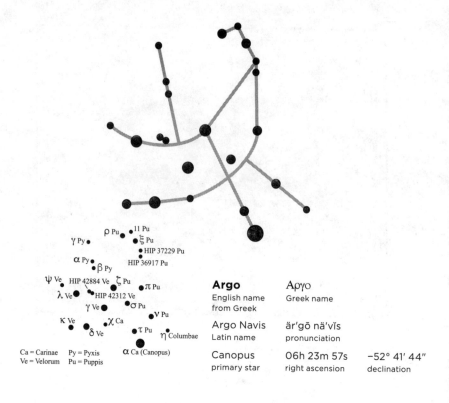

Argo
English name
from Greek

Αργο
Greek name

Argo Navis
Latin name

är′gō nä′vĭs
pronunciation

Canopus
primary star

06h 23m 57s
right ascension

−52° 41′ 44″
declination

Labels within the figure:

ρ Pu 11 Pu ξ Pu HIP 37229 Pu HIP 36917 Pu

γ Py α Py β Py

ψ Ve HIP 42884 Ve ζ Pu π Pu

λ Ve HIP 42312 Ve γ Ve σ Pu ν Pu

κ Ve χ Ca δ Ve τ Pu η Columbae

Ca = Carinae Py = Pyxis α Ca (Canopus)
Ve = Velorum Pu = Puppis

crossed the stormy seas and straits for countless days, and came at last to Hellas with their hard-earned prize. Not long later, Jason deposed his uncle, with the help of Medea, and came to reign as the rightful ruler of Thessaly. Then the Argonaut crew of heroes and friends parted to pursue further valiant adventures in their own native lands.

For his selfless service and sacrifice, the RAM received a place in heaven. Among the stars, he is seen swimming the Hellespont, with his Golden Fleece just visible above the surface of the sea. His bright muzzle turns back and up toward a guiding astral light, and his horns dip lower down.[21] Now his fleece shines for the enjoyment of every earthly being, rather than for one greedy king.

The ship ARGO also attained an eternal course in the stars,

Aries motif of the Good Shepherd appearing in Christian art. Sculpture, c. 250 AD. Courtesy of Vatican Museums, Vatican City. Photo by author.

where it navigates the celestial mists of the Milky Way. Its steering oars propel it forward, and its stern arches stately overhead. The *Argo*'s prow is concealed in the fog as it ventures into uncharted skies to the south.[22]

ORPHEUS, THE GREATEST MUSICIAN

Some of the Argonauts gained similar acclaim for their service to gods and fellow sailors while on the voyage. Orpheus—"the father of songs, and widely praised minstrel"—carried his seven-stringed Lyre aboard to offer inspiration and entertainment during the distant journey.[23] The immortals themselves had entrusted this most excellent musical instrument to him.

When Hermes—the messenger god—was but a baby, the precocious child, in search of a toy, had crafted the Lyre from a tortoise shell and a set of sacred cattle horns. Apollo was not pleased when he heard that the horns had come from one of his own beloved bulls. Hermes quickly averted his anger by handing him the sweet sounding Lyre—the first of all stringed instruments.

Apollo—the god of the sun and the arts—beamed with delight. He cherished the Lyre and carried it constantly with him. Then, one day, he gave the instrument to his true love—Calliope—as an expression of his affection.

Calliope—the Muse of epic poetry—soon discovered that the Lyre provided a perfect rhythmic companion to recitations of the stories of old. It proved to be of further worth as an accompaniment to songs, in happy harmony. So, when her son Orpheus showed an amazing aptitude for music, she passed the instrument down to him. In his talented hands, the Lyre offered a delightful addition to sacred songs and epic stories. Thus, in all the ages to follow, every bard of ability has plucked the vibrant strings of a lyre when reciting the tales of Homer and other notable poets.

With the divine instrument carefully cradled in his hands, Orpheus performed stories and songs so enchanting that he wooed the wild beasts that wandered the woods. He even "charmed the hard boulders on the mountains and the course of rivers with the sound of his songs. And the wild oak trees, signs still to this day of his singing, flourish on the Thracian shore . . . where they stand in dense, orderly rows . . . charmed by his Lyre."[24]

At the bidding of Chiron, Jason invited Orpheus to join the Argonauts. The musician immediately proved his worth. Right away, the very first day, he revealed how the Lyre offered a steady rhythm for rowers. The instrument also served to calm frayed nerves—ruffled by the rigors of the voyage.

Once, when conflict broke out between two members of the crew, Orpheus grabbed the Lyre in his gifted hands and began to sing a soothing song. "He sang of how the earth, sky, and sea, at one time combined together in a single form . . . and of how the stars and moon and paths of the sun always keep their fixed place in the sky; and how the mountains arose; and how the echoing rivers with their nymphs and all the land animals came to be."[25]

The crew—suddenly entranced in silent stares, with mouths opened in awe—watched and listened to the minstrel, and lost all thought of strife. Soon, they merrily continued their course.

On the return journey, the crew succumbed to the eerie enchantment of the sweet-voiced Sirens. These seductive, deadly sea nymphs lived on a rocky island and lured the sailors of passing ships. Their irresistible song drew many men into dangerous shoals and sent them, and their boats, to a watery grave. Fate proved even worse for those who made landfall among the "femmes fatales." In their loving, lethal embrace, men languished and slowly died.

Even the Argonauts could not resist their charms. "Already they were about to cast the cables from their ship onto the beach." But, "Orpheus . . . strung his Bistonian Lyre in his hands and rung out the rapid beat of a lively song, so that at the same time the men's ears might ring with the sound of his strumming." The Lyre promptly overpowered the song of the Sirens; and "the Zephyr and the resounding waves, rising astern, bore the ship onward"—far away to safety.[26]

At journey's end, Orpheus returned to his native land. There, a throng of impassioned women fell in love with the famous bard, as often befalls musicians and poets. When he performed and "beat the ground rapidly with his shining sandal to the accompaniment of his beautifully strummed Lyre and song," they swooned

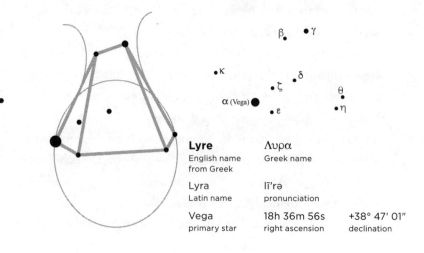

Lyre
English name
from Greek

Λυρα
Greek name

Lyra
Latin name

lī′rə
pronunciation

Vega
primary star

18h 36m 56s
right ascension

+38° 47′ 01″
declination

or became flushed in frenzy.[27] Finally, a mob of spellbound women, in jealous rage for his affection, tore him limb from limb.

Orpheus figured foremost among men as the first of mortal musicians, and as a poet and prophet. Having received his inspiration directly from the Muses, and his instrument as a gift from the gods, no human was ever able to match his musical skill. Zeus honored Orpheus—his grandson through his daughter Calliope—by allowing the LYRE a prominent place in the sky. Here it is marked by one of the brightest of stars. For, as Homer truly said so long ago: "among all men that are upon the earth minstrels win honor and reverence."[28]

ASCLEPIUS, THE FAMED PHYSICIAN

One of Orpheus' shipmates, Asclepius, served as physician and surgeon on the *Argo*. During the long and perilous voyage he preserved the health and welfare of the men. Asclepius had learned the medicinal use of plants from Chiron and carried his own herbarium close at hand.[29] To this he added flora found in foreign lands along the route, whenever their botanical properties proved of worth.

He also applied the means of healing he had learned in Hellas. As a youth, he watched the mysterious methods of slithering snakes and studied their amazing restorative arts. For ages, the wise and wily creatures had devised incredible cures—such as shedding their skins for annual rejuvenation. But they kept their knowledge closely concealed from prying eyes.

Asclepius came to respect and befriend them, and often observed their ways. He even adopted, as his medical symbol, a serpent coiled around a staff.[30] Other mortals also esteemed the wisdom of snakes and entrusted them as guardians of sacred temples, shrines, and springs. The Athenians kept a sizable snake on the crest of the Acropolis for this purpose and fed it a monthly fare of honey cake.[31]

Asclepius became so absorbed in the healing arts that he overstepped his human bounds. After his return from Colchis, he heard that Theseus' son—Hippolytus—had died without warning. To console the tragedy-stricken father, Asclepius rallied all his awesome powers and brought them to bear on Hippolytus' lifeless body. With firm persistence and presence of mind, he brought the lad back to the land of the living.

But Zeus was angry. How dare Asclepius to delve into the realm of the gods! In one impetuous moment, Zeus's thunderbolt darted down to earth and dashed away the life of the famed physician.

Then, in a fit of remorse, the flighty god commemorated Asclepius' astounding service to humankind by placing him and his favorite serpent in the sky. Asclepius appears in starry splendor as

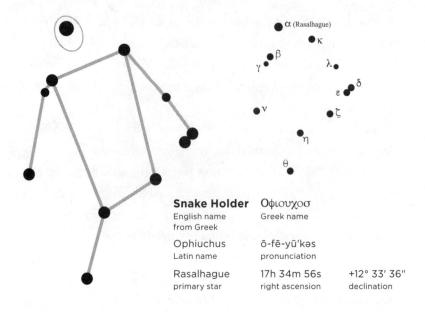

Snake Holder	Οφιουχοσ	
English name from Greek	Greek name	
Ophiuchus	ō-fē-yū′kəs	
Latin name	pronunciation	
Rasalhague	17h 34m 56s	+12° 33′ 36″
primary star	right ascension	declination

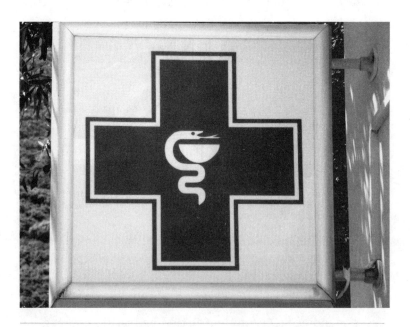

The Snake of Ophiuchus as a symbol of medicine on a modern pharmacy sign in Stavros, on the island of Ithaca, Greece. Photo by author.

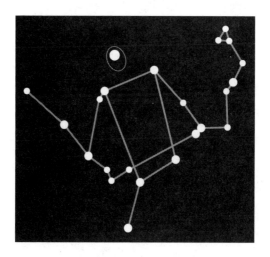

Snake Holder and Snake as they appear together in the sky.

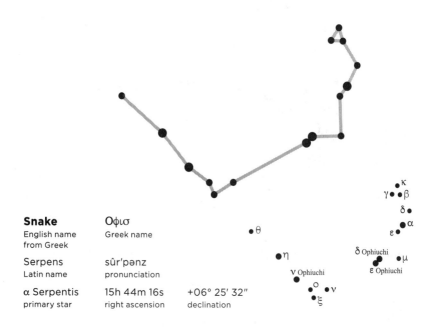

Snake
English name
from Greek

Οφισ
Greek name

Serpens
Latin name

sûr'pənz
pronunciation

α Serpentis
primary star

15h 44m 16s
right ascension

+06° 25' 32"
declination

Asclepius and his snake. Plaster cast from a marble original of the fourth century BC. Courtesy of the Museum of Epidaurus Theater, Epidaurus, Greece. Photo by author. A fragmented Roman copy, c. 160 AD, is preserved in the National Archeological Museum, Athens, Greece.

the SNAKE HOLDER and gently grasps the SNAKE in his caring hands. In this way, the symbol of medicine is forever portrayed at night. And, to this day, physicians with healing hands take the Hippocratic Oath in the name of Asclepius.

THE DEVOTED TWINS

Two more Argonauts, known as the TWINS—Castor and Polydeuces—earned a place in heaven for their lasting love for one another and for other mortals and gods. Out on the open ocean, the Twins amazed the crew with their uncanny skills of navigation. They often assisted Tiphys and Lynceus to find a favorable route.

The Argonauts also admired the Twins' divine devotion and called on them as intercessors for safety at sea. During one especially fierce and frightening storm, as the disheartened men fell on the deck in despair, the Twins "stood up and raised their hands to the immortals and prayed." Slowly, the tempest subsided, and the crew—reassured—returned to their feet.[32]

Castor and Polydeuces served the Argonauts with such devotion that Zeus "entrusted them with ships of future sailors as well."[33] From that day forward, seafarers through the ages, when seeking reassurance, have looked to the constellation of the Twins—with its two bright stars that mark their heads.[34] For added protection in open waters, mariners paint or carve icons of the two men on the bows of their ships. And when two fiery balls of light appear on the masthead toward the end of a storm at sea, reverent seamen thank the Twins for their sign of assurance that all is well.[35]

The Twins helped the Argonauts find the way back home to Hellas through the same uncanny knack that had helped them find each other as children. Whenever the toddlers wandered apart, they searched with tearful eyes until, at a distance, one spied the other and hurried to a happy reunion. As adults, they remained the best of friends and grew old together. But one day, many years after returning from Colchis, Castor failed to wake up in the morning. His soul had left his aged body and slipped away in the night.

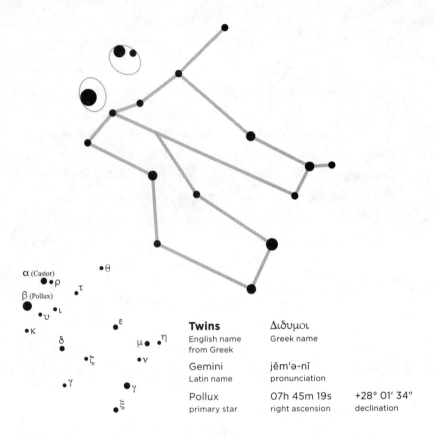

α (Castor)
β (Pollux)

κ
θ
ρ
τ
υ ι
δ
ε
μ η
ζ ν
γ
γ
ξ

Twins
English name
from Greek

Gemini
Latin name

Pollux
primary star

Διδυμοι
Greek name

jĕm′ə-nī
pronunciation

07h 45m 19s
right ascension

+28° 01′ 34″
declination

Polydeuces was lost in loneliness and grief. His vision blurred and his hearing became muffled. His mind wandered in aimless confusion, as if in a fog on a trackless sea. For days and months and years he mourned and prayed to the gods to restore his brother or take his life.

At last, Polydeuces' soul ascended to heaven in happy reunion with Castor. Among the stars they stand, side-by-side, together forever as constant companions, with arms around each other's shoulders. Polydeuces stands slightly behind and to the right of Castor; while Castor extends his left arm out, as if pointing the way through a narrow passage at sea.

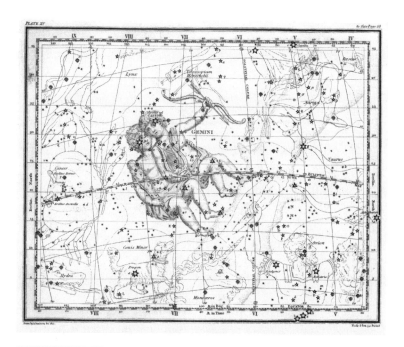

Jamieson Plate 15: Gemini

As a constellation, the Twins reflect the love and devotion they held for one another. In life, they honored the gods and served their fellow mortals—free of arrogance and greed. They represented all that Hellenes hold dear. Now, among the stars, they personify the ideals embodied in the tales of ancient skies.

PART TWO

—

THE MODERN
CONSTELLATIONS

In the late ancient, medieval, and renaissance eras, the forty-eight classical constellations continued to grace the skies. Then, around 1600 AD, the human perspective began to expand at a remarkable rate. Scientists and philosophers redoubled their rational queries into the nature of the universe.

Global exploration amassed a wealth of geographical knowledge. Technological advances produced the telescope, and other scientific instruments, to aid investigations. With this revived spirit of inquiry came renewed interest in the celestial sphere, resulting in the addition of forty-one constellations.

The voyages of the Dutch explorer Pieter Keyser took him to southern seas, beneath uncharted regions of the southern sky. Here he gazed on stars that no astronomer of ancient Greece had seen. To fill the vacant space on the celestial map, Keyser contrived twelve new constellations.

His countryman, Petrus Plancius, popularized these star

Magnitudes: -1 to 1 2 3 4

Modern Northern Hemisphere

Re = Reticulum
Ch = Chamaeleon
TA = Triangulum Australe

Magnitudes: -1 to 1 2 3 4

Modern Southern Hemisphere

groupings, and Johann Bayer portrayed them in his fabulous celestial atlas of 1603—the *Uranometria*.[1] These new constellations included Phoenix, Grus, Indus, Pavo, Triangulum Australe, Apus, Musca, Chamaeleon, Volans, Dorado, Hydrus, and Tucana. Most of these depicted exotic or mythical fauna that reflected southern latitudes.

Plancius also proposed four further star groups that later gained the rank of constellation. These included Crux, Columba, Camelopardalis and Monoceros. In the same era, the astronomer Tycho Brahe elevated the ancient asterism of Coma Berenices to the list.

A few years after the addition of these constellations, Galileo Galilei heard reports of the amazing optical properties of a Dutch magnifying toy. He soon developed a twenty-power telescope based on the concept. In November 1609, he turned his telescope toward the sky and discovered details that no human had beheld from the dawn of time.

In December, he sketched the moon's surface and confirmed the spherical shape of the planets. The following month, he found the four Galilean Moons as they slowly circled Jupiter. All the while, he improved the telescope's lenses and refined the total design.

The ensuing explosion of new knowledge led to a popular love of astronomy that remains to this day. In this milieu, scientists continued to classify poorly defined regions of the sky. Johannes Hevelius contributed seven new constellations in 1687, including Canes Venatici, Leo Minor, Lynx, Lacerta, Vulpecula, Scutum, and Sextans.

Louis de La Caille followed in 1756, with fourteen southern sky constellations: Telescopium, Norma, Antlia, Pyxis (formerly the mast of Argo Navis), Caelum, Fornax, Sculptor, Microscopium, Octans, Circinus, Pictor, Mensa, Reticulum, and Horologium. Finally, seven years later, La Caille divided the oversize classical constellation of Argo Navis into Vela (formerly the ship's deck and hull), Carina (formerly the keel and port steering oar), and Puppis (formerly the stern and upper steering oars).

Keyser, Plancius, and Hevelius continued the ancient custom of naming most of the constellations after animals. But La Caille departed from the practice. Instead, he reflected the tone of his time by marking groups of stars as mundane inventions of the scientific and industrial revolutions.

With these, the last of the eighty-eight permanent constellations appeared. Still, the elegant beauty and ancient lore of the forty-eight classical constellations commanded the night sky.

Chapter 7 illustrates the forty-one modern constellations. Each chart is labeled with the Latin name and pronunciation, the English translation, the primary star, and its right ascension and declination. Chapter 8 shows the sharp departure from the realistic forms of ancient constellations to the abstract forms of modern times.

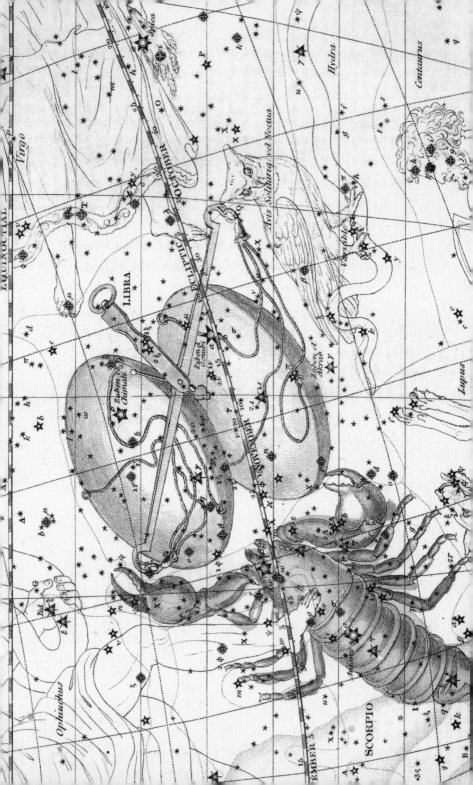

7

FILLING IN THE CELESTIAL SPHERE

PIETER KEYSER
—

PHOENIX, PAVO, GRUS, INDUS, TRIANGULUM AUSTRALE,
MUSCA, APUS, CHAMAELEON, VOLANS, HYDRUS,
DORADO, TUCANA

PETRUS PLANCIUS AND TYCHO BRAHE
—

MONOCEROS, COLUMBA, CAMELOPARDALIS, CRUX,
COMA BERENICES

JOHANNES HEVELIUS
—

CANES VENATICI, LEO MINOR, LYNX, LACERTA, VULPECULA,
SCUTUM, SEXTANS

LOUIS DE LA CAILLE
—

TELESCOPIUM, NORMA, SCULPTOR, PYXIS, FORNAX, ANTLIA,
CAELUM, MICROSCOPIUM, OCTANS, CIRCINUS, HOROLOGIUM,
MENSA, RETICULUM, PICTOR, VELA, CARINA, PUPPIS

Left: Jamieson Plate 19: Scorpius, Libra

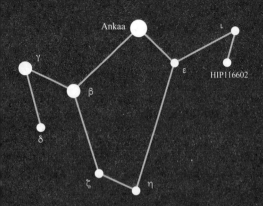

Phoenix
Latin

Phoenix	fē'nĭks	
English	pronunciation	
Ankaa	00h 26m 17s	−42° 18' 22"
primary star	right ascension	declination

Pavo
Latin

Peacock	pä'vō	
English	pronunciation	
Peacock	20h 25m 39s	−56° 44' 06"
primary star	right ascension	declination

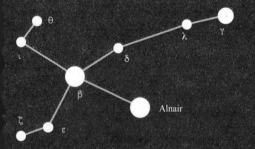

Grus
Latin

Crane	grŭs'	
English	pronunciation	
Alnair	22h 08m 14s	−46° 57' 40"
primary star	right ascension	declination

Indus
Latin

Indian	ĭn'dəs	
English	pronunciation	
α Indi	20h 37m 34s	−47° 17' 29"
primary star	right ascension	declination

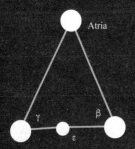

Triangulum Australe
Latin

Southern Triangle	trī-ăn′gyə-ləm ô-strā′lē	
English	pronunciation	
Atria	16h 48m 40s	−69° 01′ 40″
primary star	right ascension	declination

Musca
Latin

Fly	mŭs′kə	
English	pronunciation	
α Muscae	12h 37m 11s	−69° 08′ 08″
primary star	right ascension	declination

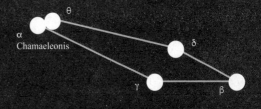

Apus
Latin

Bird of Paradise	ā′pəs	
English	pronunciation	
α Apodis	14h 47m 52s	−79° 02′ 41″
primary star	right ascension	declination

Chamaeleon
Latin

Chameleon	kə-mēl′yən	
English	pronunciation	
α Chamaeleonis	08h 18m 32s	76° 55′ 11″
primary star	right ascension	declination

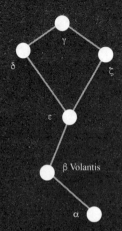

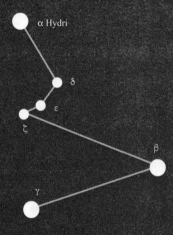

Volans
Latin

Flying Fish · vō′länz
English · pronunciation

β Volantis · 08h 25m 44s · −66° 08′ 13″
primary star · right ascension · declination

Hydrus
Latin

Water Snake · hī′drəs
English · pronunciation

α Hydri · 01h 58m 46s · −61° 34′ 11″
primary star · right ascension · declination

Dorado
Latin

Swordfish · dô-rä′dō
English · pronunciation

α Doradus · 04h 33m 60s · −55° 02′ 42″
primary star · right ascension · declination

Tucana
Latin

Toucan · tū-kä′nə
English · pronunciation

α Tucanae · 22h 18m 30s · −60° 15′ 35″
primary star · right ascension · declination

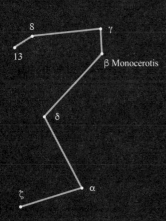

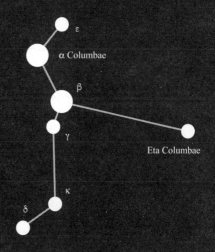

Monoceros
Latin

Unicorn
English

mə-nŏs'ə-rəs
pronunciation

β Monocerotis
primary star

06h 28m 49s
right ascension

−07° 01' 59"
declination

Columba
Latin

Dove
English

kə-lŭm'bə
pronunciation

α Columbae
primary star

05h 39m 39s
right ascension

−34° 04' 27"
declination

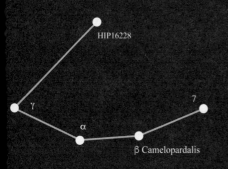

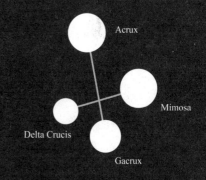

Camelopardalis
Latin

Giraffe
English

kə-mĕl'ō-pär-də-lĭs
pronunciation

β Camelopardalis
primary star

05h 03m 25s
right ascension

+60° 26' 32"
declination

Crux
Latin

Southern
Cross
English

krŭks'
pronunciation

Acrux
primary star

12h 26m 36s
right ascension

−63°,05' 57"
declination

β Comae Berenices

γ

α

β

α Canum Venaticorum

Coma Berenices
Latin

Berenice's Hair	kō′mə bĕr-ə-nī′sēz	
English	pronunciation	
β Comae Berenices	13h 11m 52s	+27° 52′ 41″
primary star	right ascension	declination

Canes Venatici
Latin

Hunting Dogs	kā′nēz vĭ-năt′ə-sī	
English	pronunciation	
α Canum Venaticorum	12h 56m 02s	+38° 19m 06s
primary star	right ascension	declination

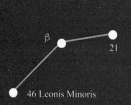

β

21

46 Leonis Minoris

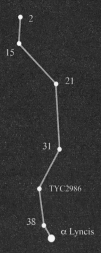

2

15

21

31

TYC2986

38

α Lyncis

Leo Minor
Latin

Little Lion	lē′ō mī′nər	
English	pronunciation	
Leonis Minoris	10h 53m 19s	+34° 12′ 54″
primary star	right ascension	declination

Lynx
Latin

Lynx	lĭngks′	
English	pronunciation	
α Lyncis	09h 21m 03s	+34° 23′ 33″
primary star	right ascension	declination

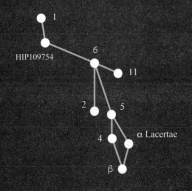

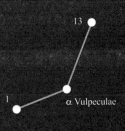

Lacerta
Latin

Lizard
English

lə-sûr′tə
pronunciation

α Lacertae
primary star

22h 31m 18s
right ascension

+50° 16′ 57″
declination

Vulpecula
Latin

Little Fox
English

vŭl-pĕk′yə-lə
pronunciation

α Vulpeculae
primary star

19h 28m 42s
right ascension

+24° 39′ 54″
declination

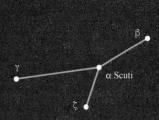

Scutum
Latin

Shield
English

skū′təm
pronunciation

α Scuti
primary star

18h 35m 12s
right ascension

−08° 14′ 39″
declination

Sextans
Latin

Sextant
English

sĕks′təns
pronunciation

α Sextantis
primary star

10h 07m 56s
right ascension

−00° 22′ 18″
declination

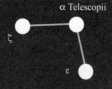

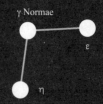

Telescopium
Latin

Telescope	tĕl-ə-skō′pē-əm	
English	pronunciation	
α Telescopii	18h 26m 58s	−45° 58′ 06″
primary star	right ascension	declination

Norma
Latin

Level	nôr′mə	
English	pronunciation	
γ Normae	16h 19m 50s	−50° 09′ 20″
primary star	right ascension	declination

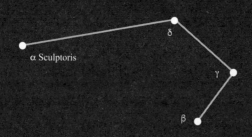

Sculptor
Latin

Sculptor	skŭlp′tər	
English	pronunciation	
α Sculptoris	00h 58m 36s	−29° 21′ 27″
primary star	right ascension	declination

β

α Pyxidis

γ

Pyxis
Latin

Compass (of ship)	pĭk′sĭs
English	pronunciation

α Pyxidis
primary star

08h 43m 36s
right ascension

−33° 11′ 11″
declination

ν

α Fornacis

β

Fornax
Latin

Furnace	fôr′năks
English	pronunciation

α Fornacis
primary star

03h 12m 05s
right ascension

−28° 59′ 15″
declination

ε

θ

ι

α Antliae

Antlia
Latin

Pump	ănt′lē-ə
English	pronunciation

α Antliae
primary star

10h 27m 09s
right ascension

−31° 04′ 04″
declination

β

α Caeli

Caelum
Latin

Chisel	sē′ləm
English	pronunciation

α Caeli
primary star

04h 40m 34s
right ascension

−41° 51′ 50″
declination

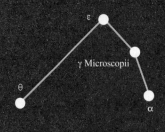

Microscopium
Latin

Microscope	mīk-rə-skō'pē-əm	
English	pronunciation	
γ Microscopii	21h 01m 17s	−32° 15′ 28″
primary star	right ascension	declination

Octans
Latin

Octant	ŏk'tănz	
English	pronunciation	
ν Octantis	21h 41m 29s	−77° 23′ 24″
primary star	right ascension	declination

Circinus
Latin

Compass (for drawing)	sîr'sə-nəs	
English	pronunciation	
α Circini	4h 42m 30s	−64° 58′ 30″
primary star	right ascension	declination

Horologium
Latin

Clock	hôr-ə-lō'jē-əm	
English	pronunciation	
α Horologii	04h 14m 00s	−42° 17′ 40″
primary star	right ascension	declination

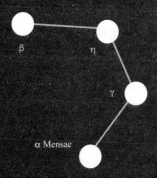

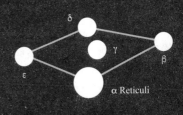

Mensa
Latin

Table
English

mĕn'sə
pronunciation

α Mensae
primary star

06h 10m 14s
right ascension

−74° 45′ 11″
declination

Reticulum
Latin

Net
English

rĕ-tĭk'yə-ləm
pronunciation

α Reticuli
primary star

04h 14m 25s
right ascension

−62° 28′ 26″
declination

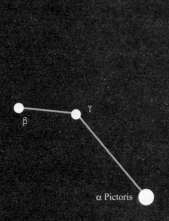

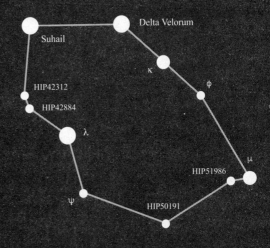

Pictor
Latin

Painter's
Easel
English

pĭk'tôr

pronunciation

α Pictoris
primary star

06h 48m 11s
right ascension

−61° 56′ 29″
declination

Vela
Latin

Sail (of ship)
English

vē'lə
pronunciation

Suhail
primary star

08h 09m 32s
right ascension

−47° 20′ 12″
declination

Carina
Latin

Keel (of ship)   kə-rī′nə
English   pronunciation

Canopus   06h 23m 57s   −52° 41′ 44″
primary star   right ascension   declination

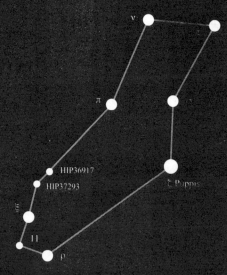

Puppis
Latin

Stern (of ship)   pŭp′ĭs
English   pronunciation

ζ Puppis   08h 03m 35s   −40° 00′ 12″
primary star   right ascension   declination

8

MODERN
MISREPRESENTATIONS
OF THE ANCIENT
CONSTELLATIONS

In the eighteenth century, the astronomer Louis de La Caille devised new constellations for the southern skies and named them after mundane objects. In the same era, astronomy became more strictly scientific and clinical. It lost much of the humanity that had characterized earlier celestial observations.

Greek mythology had breathed life into the shining constellations—animating the skies with personality and meaning. Ancient astronomers fondly preserved these images of old, even while carefully conducting scientific investigations. But in modern times, more often than not, these images became replaced by lifeless abstractions. The following diagrams illustrate this change by showing identical star groups, with typical modern forms above and ancient forms below.

To add to the problem, recent attempts to depict the constellations according to their ancient designations have fallen short— failing to reflect the original images. The hope here is to see the ancient stars and stories shine again, alongside the scientific study of astronomy. In this way, the boundless beauty and timeless tales of the forty-eight classical constellations may remain, even as they persisted for three millennia.

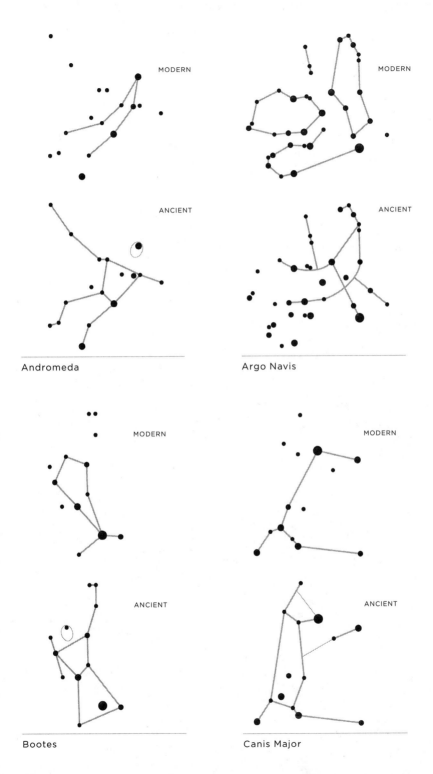

MODERN

ANCIENT

Andromeda

MODERN

ANCIENT

Argo Navis

MODERN

ANCIENT

Bootes

MODERN

ANCIENT

Canis Major

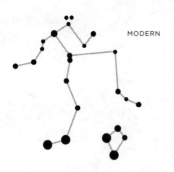

MODERN

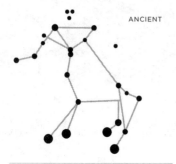

ANCIENT

Centaurus

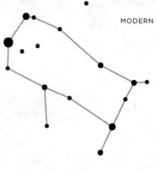

MODERN

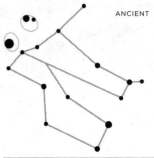

ANCIENT

Gemini

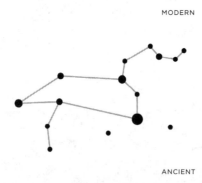

MODERN

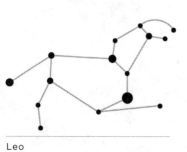

ANCIENT

Leo

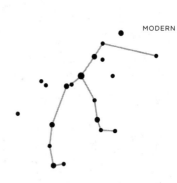

MODERN

ANCIENT

Perseus

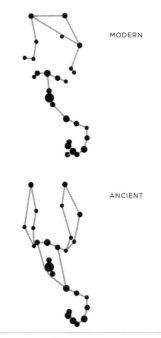

MODERN

ANCIENT

Scorpius-Libra

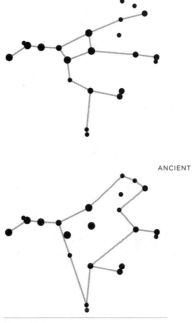

MODERN

ANCIENT

Ursa Major

PART THREE

FARMING, SHEPHERDING, AND SAILING BY THE STARS

9

THE ANCIENT CELESTIAL CALENDAR

Throughout the year, "as the seasons rolled on . . . and the many days were brought in their course," the Greeks looked to the stars to faithfully measure the passage of time.[1] In the eighth century BC, Homer mentioned the major segments of the year as seasons and months. Later astronomers defined the "twelve signs of the Zodiac" that "mark the great year—the season to plow and sow the fallow field and the season to plant the tree."[2]

Ancient farmers depended on these and other celestial signposts while working to feed their families. Star and constellation risings and settings told them when to plow and plant, and when to prune the grapevines. They told them when to harvest and winnow the grain, and when to gather the grapes.

Anxious eyes turned toward the skies and closely scanned the eastern horizon before the light of morning. As dimmer stars disappeared with the approach of dawn, observers focused all the more closely. They watched and waited for certain bright stars to make their heliacal rising—their first annual appearance in the east ahead of the sun.[3]

As farmers became more familiar with the patterns of stars and constellations, they learned to predict specific risings. They knew, for example, that the star called Herald of the Dog (Procyon) rose eight days ahead of the Dog Star (Sirius). They knew that by noting a star's position in its constellation, they could predict the star's approach as the constellation inched higher into the eastern sky each morning. Likewise, if a star could not be seen because it was "dark with clouds or hidden behind a hill," they could use surrounding stars to surmise its location.[4]

Farmers soon became proficient at assigning principal stars and constellations to certain seasons. In ancient times, the winter months closely matched our modern calendar dates of December 27 to March 28.[5] During this three-month period, the Greeks discerned that the sun passed through the three zodiacal constellations associated with water—Goat Horn (Capricornus), Water Bearer (Aquarius), and Fishes (Pisces). These marked the cold and rainy season, when fields remained mostly unattended. During this time, farm families sat by the fire and busied themselves with indoor chores, such as repairing and replenishing tools and furnishings.

The wood for these had been chopped in autumn—at the end of the harvest of grapes and olives—when trees had entered dormancy and dropped their leaves. Autumn was also the time to fashion new plows, if needed. One was for the fall plowing at hand. A second served as a spare in case the first should break. The sturdy wood of the holm oak tree produced the best plows—with a hand-width trunk for a running board and an angled branch for a handle.[6]

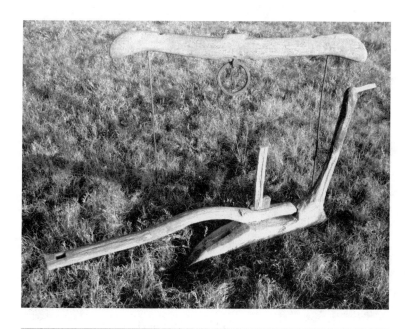

Oak plow and yoke typical of the ancient Mediterranean. Marshall Collection, Loma Paloma, Texas. Photo by author.

Other toolmaking could wait for winter, when farmers stayed inside fashioning handles, sharpening iron, and attaching the two together to make the tools. The tools included mattocks for digging and forming furrows, hoes for weeding and working the soil, sickles for harvesting stalks of grain, and knives for pruning vines and shoots. Some farmers crafted two-wheeled carts for toting tools, fodder, and fertilizer. At harvest, the carts hauled hefty loads of grain, grapes, and olives. Farmers who owned no grinding stone might also make a wooden mortar and pestle—for pounding parched grain into meal.[7] Carving, shaping, and smoothing tools and furnishings kept families busy through the long winter months.

Finally, by February 25, "sixty wintry days after the solstice," the coldest months had come and gone. Now, after sunset, farmers stared into the eastern sky, anticipating the appearance of the star they called Bear Guard (Arcturus). They knew that it would soon

Ancient Greek-Scythian knife and farm tools typical of the ancient Mediterranean: sickle, axe, hoe, and mattock. Marshall Collection, Loma Paloma, Texas. Photo by author.

rise over the sparkling sea, "shining brightly just at dusk, leaving behind the holy stream of Oceanus."[8] On the day following the star's first appearance, they left their cozy hearths and homes, and headed for the fields with freshly sharpened knives and mattocks.[9]

The time had arrived to prune the vineyards and plant the fruit trees—figs, apples, pears, and olives.[10] For planting, the farmer first selected a site free of saturation. Too much water would rot the roots. Then he dug a hole no more than two feet wide and two and a half feet deep.

He left a loose layer of loam in the bottom to give the roots ample room to grow and roam. As he planted the tree and filled the hole, he firmly packed soil around the sapling. This blocked the sun from baking the earth and wilting the roots, and barred the rain from forming a bog.[11]

Although the star named Bear Guard promised the approach of spring, the skies might still bring fitful bursts of wet or wintry weather. For farmers, weather made the difference in survival or starvation. So they learned to read the signs of coming rain or drought; or scorching heat; or frost and freeze; or heavy winds and storms. Across the generations, they amassed a wealth of weather lore.

They noted that a soft, rosy dawn brought an interval of calm. But when the morning sky shone fiery red or waterfowl wheeled overhead and fled the sea, then one was wise to watch for a coming storm. For further signs of rain, they observed the nervous chatter of sparrows at early light or swarms of bees and wasps throughout the day. Or they noticed wandering toads and salamanders; or the shrill songs of frogs; or the sputter and spark of an oil lamp that proved hard to light.[12]

On March 28, a month after the sighting of Bear Guard, the sun entered the constellation of the Ram (Aries) to mark the start of spring. This was vernal equinox, when days and nights were equal in duration. Farmers now found their way into the fields to hoe and chop the weeds; and thin the overcrowded stalks of grain; and cover roots exposed by wind and rain.[13]

According to ancient Greek legend, the time had come for poor Persephone to return from six somber months among the dead—from the dark, dreadful halls of Hades. For six more months to follow, she happily rejoined her mother—Demeter—the goddess of farming and harvest. With Demeter's sorrow for her daughter set aside, winter changed to spring; the gloomy earth turned green; and swallows arrived to fill the air with their sweet, lisping lilts and graceful aerobatics.[14]

In the hearts and minds of Greeks, swallows represented spring and figured foremost among their feathered friends. According to Aelian: "A swallow is a sign that the best season of the year is at hand. And it is friendly to man and takes pleasure in sharing the same roof. . . . Men welcome it in accordance with the law of hospitality laid down by Homer, who bids us cherish a guest while he is with us, and speed him on his way when he wishes to leave."[15] Kites and other migratory birds also served as heralds of spring, in the same manner that their departure would later mark the coming winter.[16]

Several days into spring, on April 5, farmers beheld the heliacal setting of the Pleiades. As they stood in the deepening dusk, they watched the last appearance of the shining asterism as it followed the setting sun beyond the western horizon. After this cosmic event, they counted forty days while waiting for the Pleiades to return. To pass the time, they sharpened tools in preparation for the toilsome tasks ahead—the reaping, threshing, and winnowing of barley and wheat.

At last, on May 16, the Pleiades appeared in the dim light of dawn—making a heliacal rising ahead of the morning sun. Now, all those who could hold a sickle in hand headed for the field to

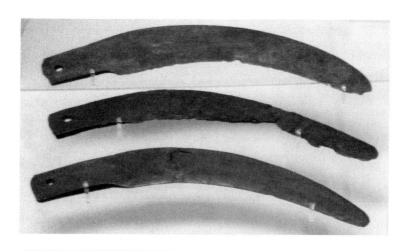

Sickles used in the grain harvest, c. 1250–1180 BC. Courtesy of the Archeological Museum, Mycenae, Greece. Photo by author.

start the harvest.[17] With backs turned to the wind, to shield the eyes from dust and debris, the strenuous task of cutting stalks of grain began. If the reaper saw that the grain stalk was short, he cut it low. If the stalk was tall, he cut it high. This provided the same length of stalk for fodder, but not so much that it hampered threshing and winnowing.

The stubble that stayed in the field was later burned to enrich the soil or tossed upon a pile for compost.Otherwise, the farmer stacked portable pens to allow livestock to graze the stubble. In return, the animals left a rich layer of manure that fertilized the field.

After cutting stalks of grain, reapers bundled and hauled them to the threshing floor. When fields yielded large crops, the harvest could continue for more than a month. Farmers hoped to gather all the grain by the time the star named Betelgeuse, in the constellation of Orion, made its heliacal rising on June 29. This also marked the time that the Crab (Cancer) aligned with the sun on summer solstice.

Grain harvest of Mycenae. Fresco, c. 1250–1180 BC. Courtesy of the Archeological Museum, Mycenae, Greece. Photo by author.

Orion's arrival ushered in the season for threshing and winnowing. Laborers now laid their stalks of grain on a patch of hard-packed earth, "in a well-aired place . . . on a well-rolled threshing floor."[18] At the center of the floor stood a stout vertical post. To this they tied a yoked pair of oxen, horses, mules, or donkeys. In response to shouts and prods, the beasts walked in slow circles around the post, treading the stalks with hard hooves to separate husks from seeds. As the animals lumbered along, threshers watched for untouched stalks and threw them under their trampling feet.[19]

Threshing took place on a site exposed to the breeze so that grain could be winnowed on the same spot.[20] After leading the weary livestock away, the workers removed the larger scraps of stalk from the ground. Then, beginning on the downwind half of the threshing floor, they turned their backs to the wind and tossed the trampled plants into the air.

As the wind blew the chaff away, the heavier seed fell to the floor. Swiftly the winnowers swept the precious grain and placed it into pots or baskets. Then they repeated the process on the upwind half of the floor. At last, with the seed safely stored, they bundled the stalks for fodder.[21]

Each morning, as the workers watched Orion rising higher in the sky ahead of the sun, they rushed to complete their tasks on time. Often the threshing and winnowing continued from dawn to dusk. They knew that a delay might drag the labor into the sweltering heat of summer, when itchy chaff stuck to sweaty skin, and fatigue began to take its toll.

When the star called Herald of the Dog (Procyon) appeared on the morning of July 20, farmers knew that the Dog Star (Sirius) could not be far away. Sure enough, after a wait of eight days, this brightest star in the sky rose at dawn to announce the scorching Dog Days of summer. Following close behind, the sun entered the constellation of the Lion (Leo).

The Dog Star shone so bright that many believed its intense light made the Earth hotter for the following fifty days. The Greeks recalled how their ancestors had prayed to Zeus for relief from the sweltering heat. Zeus, in turn, commanded the northern (Etesian) winds to blow for forty days each year to make the summers somewhat less severe.[22]

Still, the Dog Days remained too hot for heavy labor. Even the ancient Greek poet Hesiod—the hardworking farmer—advised a rest. He urged that when the cicada "pours forth its voice in the most dread heat," then is the time to sit in the shade and drink wine mixed with cold spring water, while nibbling bread, cheese, and meat.[23]

At last, as the blazing Dog Days came to a close, the grape harvest loomed ahead as a highlight of the year. Over the past several months, farmers had tended their vineyards whenever time allowed, from the late-winter pruning until now. Xenophon noted how Nature herself had taught men the means of assisting the vines throughout the year.

For example, Nature showed humans that vines liked to climb, so gardeners indulged them by erecting trellises. Later, when young and tender clusters of infant grapes emerged, the vines promptly pampered them by spreading leaves to save them from the scorching sun. Responding in kind, workers added man-made shades to keep them safe. At last, when the grapes grew large and longed to ripen in the sun, the vines started to shed their leaves. So farmers assisted by stripping remaining foliage.[24]

On September 13, the star called Herald of the Vintage (Vindemiatrix) rose at dawn to announce the long-anticipated time to gather the grapes. The Bear Guard (Arcturus) followed eight days later to hasten all remaining hands to the harvest. During these first delightful days of fall, the favored fruit was happily picked to the lively tune of the flute.

The best vineyards grew "clusters of all sorts . . . which ripened one by one at separate times." Such was the case at the lovely farm of Laertes—the father of Odysseus—on the fertile island of Ithaca.[25] With vineyards like these, the pickers made multiple passes to pluck the fruit as soon as it ripened.[26]

Now came the time to follow Hesiod's method of making wine: "Pluck off all the grapes and take them home. Set them out in the sun for ten days and ten nights, then cover them up in the shade for five, and on the sixth, draw out the gift of much-cheering Dionysus [god of wine and revelry] into storage-vessels."[27]

Owners of large vineyards paid pickers to stomp the grapes in vats, again to the rhythm of flutes. Families with smaller harvests often pressed their grapes by hand. With the work at an end, and the wine safely stored in pottery or leather skins, farmers ceremoniously sampled the vintage. Meanwhile, autumn arrived as the sun entered the constellation of the Claws (Libra), on September 30—at autumnal equinox.

By now, the olive harvest beckoned. It began earlier or later in some locales, depending on climate, desired degree of ripeness, and other considerations. Harvesters picked the olives by hand from the lower branches and knocked down those above with the gentle rap of a reed. Some they kept for eating. Others, they pressed to make oil for cooking and lighting lamps.

As the olive harvest progressed, the Pleiades dipped lower each night toward the west. Finally, the shimmering asterism reached its apparent cosmical setting—its last appearance on the western horizon just before the eastern sunrise. This celestial event, on November 3, marked the time for plowing and sowing the grain-fields, before the onset of winter.

Following the Pleiades, the Hyades and Orion also set in the west, on November 11 and 25. Together, these provided three celestial warnings for farmers to finish their plowing before foul weather arrived.[28] If you wait too long, warned Hesiod, and "plow the divine earth first at the winter solstice, you will harvest sitting down . . . grasping only a little with your hand."[29]

Other signs of nature also urged farmers forward. They knew that when the mastic tree bore fruit, the field awaited plowing. And, when wasps swarmed all around, winter cold was coming.[30]

But the behavior of migrating birds served as the best barometer. Birds with a short range of flight, like quail, began their southern journey early in the fall. Stronger birds with a longer range, such as cranes, left later. From this, farmers could follow the flights of separate species and predict the approach of winter.[31]

They knew that, when the migratory season seemed short and swift, with cranes leaving early in large flocks that filled the sky, then winter weather was rushing down upon them. But when the cranes dallied, and departed in small, scattered groups, then cold weather was slow in coming. As a general rule, when one could hear the crying flocks of cranes, the time drew near to plow and sow the field.[32]

As farmers prepared for this major event in the annual agrarian cycle, they hauled their plows, tools, and teams of oxen or equines into place. They laid the yoke atop their beasts of burden and tied it to their necks or horns. Then, with leather straps, they attached the yoke to the plow by means of a pole.[33]

Farmers prized, most of all, a well-matched team of oxen for plowing, threshing, and dropping manure to fertilize the fields.[34] Hesiod declared that a pair of nine-year-old oxen and a forty-year-old man made the best match for plowing. This stood to reason because, being mature, they paid attention to their work, without being too old for the toil.[35]

Plowing took place when the soil held sufficient moisture for sowing seeds.[36] Ideally, the ground proved damp enough for planting when the Pleiades set in the west. If not, the farmer waited for rain, as patiently as possible, while he watched the Hyades and Orion slowly descend toward the western horizon.[37] If the rains came late, he could only hope for a delay of winter weather as well.

When at last the time was right, the plowman prodded his team to form the first furrow. In one hand, he held the handle of the plow. In the other he brandished a stick for prompting the animals ahead and for scraping the moist soil from the plowshare. Behind the plowman walked a worker who used a mattock or hoe to shape the furrows and to break up bulky clods.

The sower followed last in line to scatter the seeds—either thinly spread in thin or weaker soil, or thickly dispensed in thick or richer soil. The sower knew what the soil could bear and had the knack for steadily spreading seeds. As Xenophon said, "Sowers, no less than lyre players, need practice."[38]

With a rhythmic kick of the soil, the sower also covered the seeds. This kept the birds from filling their craws and decreasing the size of the crop. For further protection, farmers called on dogs to harass hungry flocks of birds. In later months, dogs also defended the tender shoots of barley and wheat from hares and other grazing animals.[39]

With winter grains now in the ground, germination and slow growth began. In mild winters, the plants continued to grow below the soil. But in bitter cold, the life cycle remained on hold until warmer days arrived. Meanwhile, winter weather delivered moisture that made the grain grow strong and promised a healthy harvest.

Thus proceeded the annual agrarian pattern—from plowing and sowing in November, to the harvest of grains, grapes, and olives from May to the next November. Most farmers in the Mediterranean basin followed the same cycle.[40]

While a farmer subjected one field to plowing, planting, hoeing, and harvest, a second field—that was worked the previous year—now remained at rest and fallow. This ensured the lasting fertility of the fields. Without allowing for fallow fields, soils lost their nutrients and fell into fatigue, followed by crop failure. Hesiod observed that, "Fallow land is an averter of death; a soother of children," because it keeps food on the table in future years.[41]

In spring, whenever time allowed, the farmer plowed the fallow field or broke it with a mattock. This killed the weeds before they sprouted seeds and turned them under to make a fertile mulch. In summer, the farmer furrowed the field again to uproot weeds and leave them withering in the sun.[42] When autumn arrived, the farmer plowed and planted the fallow field while allowing the recently harvested field to rest through the coming year.

Hesiod urged that autumn plowing and sowing be completed as early as possible—well before the arrival of winter. In much the same way, he urged an early start to every day. A person is better off, he said, "getting up at sunrise, so that your means of life will be sufficient. For . . . dawn gives you a head start on the road; gives you a head start on your work too." Thus, the diligent person could avoid the afternoon heat of summer and the evening wind and rain during the wetter winter months.[43]

By the time the star called Eagle (Altair)—or Storm Bird—reached its heliacal rising, on December 13, winter was afoot, in full force and fury. A half month later, on December 27, the sun entered the constellation called Goat Horn (Capricornus) to mark the winter solstice. This denoted the shortest day of the year and heralded colder months ahead. After solstice, sixty wintry days would pass before the Bear Guard (Arcturus) once again arrived to announce the time of grapevine pruning, tree planting, and the approach of spring.

Farm families relied on these risings and settings of stars to grow successful crops and to survive. The rhythm of their lives reflected the seasonal cycle conveyed in the constellations. For these heavenly lights, and the gifts of grain, grapes, and olives, the Greeks gave thanks to the gods.[44]

Shepherds shared the same sentiment and followed the same celestial calendar. They also observed the migrating birds as signs of changing seasons. When the swallows and kites arrived, shepherds knew it was "time to sheer the sheep's spring wool."[45] From then until the Pleiades appeared in the morning sky, these mobile people packed their paltry possessions and marched their flocks to distant pastures.[46]

There they might reside for several months, "from spring to the rising of Arcturus" on September 21. Afterward, they brought the flocks back home for the fall and winter. As snowstorms descended, shepherds drove their sheep and goats into caves or cozy shelters.[47]

Like farmers, shepherds spent their lives outdoors and learned to predict the weather in practical ways. They lived long days and months with their sheep and goats. Thus, they knew each one of them well—by appearance and personality—and learned to determine the weather based on their behavior.

Shepherds surmised fair weather when lambs and kids grew frisky, and frolicked and played. They watched for rain when goats, let loose from shelter, rushed to feed on their fodder, then hurried to return again to the pen. They knew to expect a storm when sheep dug at the ground with their hooves, or goats huddled close together.[48]

Shepherds also enjoyed a fond familiarity with the stars and constellations. As they spent countless lonely nights beneath the skies, they cherished the stars as comforting signs. Most of all, they revered Hesperos—the evening star—and called it "the herdsmen's star." This is because it appeared in the western sky with the approach of night and led shepherds safely to their shelters.[49] Sappho—the famed poetess of Lesbos—praised Hesperos as the "fairest of all stars," that reunites "everything that shining Dawn scattered." She said of the star, "You bring the sheep; you bring the goat; you bring back the child to its mother."[50]

Shepherds, as reclusive and reflective people, often gained a reputation for reverent humility and devotion to the gods. Many Greeks believed that shepherds—in their natural surroundings—enjoyed closer communion with deities and spirits, and sometimes gained the gift of divine inspiration. Such was the case with Hesiod, the rustic shepherd and farmer, who achieved wide renown as a poet and philosopher.

The site where Hesiod received his inspiration—on the lonely slopes of Mount Helicon—also attained lasting fame. Through the ages, the place was known as the home of the Muses. In time, the Muses' mountain became home to the world's first museum—the Mouseion—where throngs of devout pilgrims erected altars, sculptures, porticos, and theaters to honor the nine daughters of Zeus.[51]

Hesiod and his fellow shepherds seemed to enjoy a close connection with heaven. Homer expressed this sentiment, felt by many countrymen, when he described a pastoral scene in the *Iliad*: "In the sky about the gleaming moon the stars shine clear when the air is windless, and into view come all mountain peaks and high headlands and glades, and from heaven breaks open the infinite air, and all the stars are seen, and the shepherd rejoices in his heart."[52]

Seafarers enjoyed a similar close relation with the stars and constellations. Their celestial observations also served the practical purpose of defining safe seasons for sailing. Some sailed in spring, at the rising of the Pleiades. Theocritus said that such was the case with the Argonauts, who took advantage of a wind from the south—a remnant of the prevailing winter pattern—to sail northeast toward the Hellespont.[53] But most mariners awaited the safe sailing season marked by the summer solstice.

Hesiod noted that "sailing is in good season for mortals for fifty days after the solstice [from June 30 to August 19], when the summer goes to its end, during the toilsome season. You will not wreck your boat then, nor will the sea drown your men."[54] The winds remained mostly moderate and steady through this time. Still, Eudoxus warned that the arrival of the brisk Etesian winds, on July 30, made sailing somewhat tricky.

Conditions declined as autumn winds approached. As the Pleiades set in the west on November 3, the sailing season came to a close. Hesiod warned: "When the Pleiades, fleeing Orion's mighty strength, fall into the murky sea, at that time blasts of all sorts of winds rage; do not keep your boat any longer in the wine-dark sea."[55]

Orion—pursuing the Pleiades—plunged below the western horizon twenty-two days later and served as a second sign to seek a harbor ahead of autumn storms. Any crew that continued sailing after this late date had best beware. The formidable winds of fall and winter sometimes shredded sails, or mangled masts and ripped them away from their stays, leaving the ship adrift on the dreadful sea.[56]

By the time the star called Storm Bird rose, on the morning of December 13, furious winter weather roared across the white-capped waters. Fourteen days later, winter solstice sent a final warning to run the ship ashore. Aratus recalled the wrathful winter tempests: "Then is the frost from heaven hard on the benumbed sailor. . . . The sea ever grows dark beneath the keels, and, like to diving seagulls, we often sit, spying out the deep from our ship, with faces turned to the shore," longing for a safe harbor.[57]

Most sailors had already stored their boats for the season, long before the arrival of winter solstice. In doing this, they heeded the words of Hesiod:

> Draw up your boat onto the land and prop it up with stones, surrounding it on all sides, so that they can resist the strength of the winds that blow moist; and draw out the bilge-plug, so that Zeus' rain does not rot it. Lay up all the gear well prepared in your house after you have folded the sea-crossing boat's wings [sails] in good order; and hang up the well-worked rudder above the smoke.[58]

Like the farmer and shepherd, the sailor now sat at home, warming his hands at the hearth and waiting out the raging winter storms. To pass the time, he tended to indoor chores and dreamed of the coming spring.

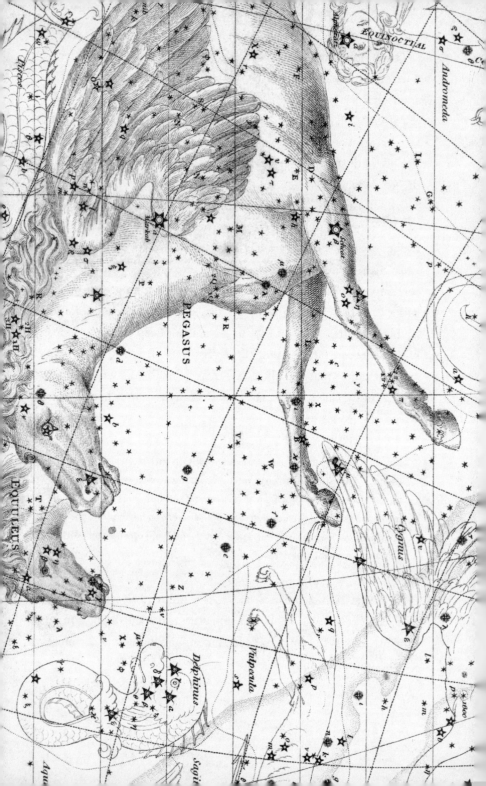

10

ANCIENT NAVIGATION

Mariners of the Mediterranean Sea sailed in some of the mildest conditions in the world. Weather patterns remained fairly stable during the standard sailing season, and waves ranged from medium to small. This held true especially in the Aegean Sea, where plentiful islands and promontories disrupted large swells and waves like those of the ocean. Tides and currents also flowed within reasonable ranges and followed familiar patterns.[1]

Even so, ancient seafaring was a dangerous proposition, as scattered shipwrecks and written records attest. Many mariners confined their cruises to the long-established safe sailing season: from summer solstice through the following fifty days (June 30 to August 19). This and other celestial signs prescribed certain times of the year to sail or stay ashore. But seamen also paid close heed to constantly changing climactic conditions.

If signs in nature favored several days of fair sailing, then seasoned sailors sometimes launched their ships right away, regardless of the time of year. For them, spring often offered fine times

Sounding weight from a shipwreck off Antikythera, Greece. Courtesy of the National Archeological Museum, Athens, Greece. Photo by author.

for sailing. Even Hesiod, the landlubber who held little love for the sea, noted that: "There is another sailing for human beings, in the springtime: at that time—when a man thinks that the leaves at the top of the fig tree are as big as the footprint a crow leaves as it goes—the sea can first be embarked upon." [2]

To sail safely, navigators needed to know how to recognize seasonal climactic changes. For this, they listened to the familiar lore of farmers and shepherds. Like them, sailors praised the swallow as a favorite sign of spring:

> It is the season for sailing; already the chattering swallow has come, and the pleasant zephyr, and the meadows' bloom, and the sea with its boiling waves lashed by the rough winds has sunk to silence. Weigh the anchors and loose the hawsers, mariner, and sail with every stitch of canvas set. [3]

As with the swallow, the arrival of the zephyr—the west wind—served as a sign that rough and rainy winter weather was at an end. The zephyr blew as a peaceful breeze in spring and signaled fair skies for sailing: "The storms have taken flight and tenderly laughing zephyr now makes the blue wave gentle as a girl." [4]

Prevailing wind and weather patterns, such as the springtime zephyr, allowed sailors to perceive a seasonal change, even when far from land. At the end of spring, the zephyr stepped aside as warm weather and stable, steady winds from the north signaled the summer season. In sharp contrast to winter, when clouds and rainy weather rolled in from the south, summer brought blue skies, and dry days and nights.

Stiff summer breezes, blowing from the north, offered fine conditions for a swift, four or five-day sail from Crete to Egypt. This route remained in use from the time of Homer, or before, and for many centuries to follow. [5] Homer's *Odyssey* described the voyage:

We embarked and set sail from broad Crete, with the North Wind blowing fresh and fair, and ran on easily as if downstream. No harm came to any of my ships, but unscathed and free from disease we sat, and the wind and the helmsman guided the ships. On the fifth day we came to fair-flowing Aegyptus, and in the river Aegyptus I moored my curved ships.[6]

North winds made this a favored route. But the return trip required a ship to hug the Asian shore for shelter from the wind. After arriving at Rhodes, the ship island-hopped toward home across the Aegean Sea.

In all parts of the eastern Mediterranean, the northern breezes of summer often blew so strong that seamen sailed off southern shores to block the blasts. The Euripus Strait, between Attica and Euboea, provided a preferable route as protection from the north.[7] But even with the stiff winds, summer remained the best time to sail.

Summer finally surrendered to fall, with its variable winds and low-pressure systems. This weather change served as a sign of winter storms ahead, when sailing became most dangerous. Flocks of cranes, in their typical faithful fashion, offered a second winter warning as they fled south.

Aristotle described how "cranes ... move [north to south] from the Scythian plains to the marshes above Egypt. . . . The fishes too migrate in the same way ... in winter from the deep sea towards the land in pursuit of the warmth, and in summer from inshore to the deep sea."[8] Aristophanes added: "When the crane whoops off to Africa [for winter] . . . it tells the ship owner to hang up his rudder."[9] Cranes and fishes both warned sailors far from shore that winter weather was afoot.

Theocritus offered a eulogy that addressed the approaching danger: "Mortal, take care of your life and do not go to sea at the wrong time; life is short as it is. Poor Cleonicus, as a merchant . . .

you were in a hurry to reach fair Thasos ... and as you made your voyage just before the setting of the Pleiads [in November], your own star set with the Pleiads themselves."[10]

Many experienced sailors, especially merchants like Cleonicus, cast caution to the wind and sailed whenever the risk seemed reasonable.[11] In fact, winter offered an opportunity for dauntless seamen to catch the south winds and cruise from the Hellespont to the Bosporus, or from Egypt back to Crete. Some mariners also sailed during the "Halcyon Days"—a half month stretch of supposed calm at the time of the winter solstice.[12]

Regardless of the season, unexpected storms and squalls might arise at any time to make maritime travel hazardous. Even in summer the experienced sailor paid close attention to weather signs. His life, and that of the crew, relied on it.

Aratus declared that the wise and wary navigator always took precautions. Before going to sleep in the evening, he furled his sails "for fear of the morning sea."[13] For no one could know what weather might arise at night to wreak destruction by morning.

Homer described one frightful night that Odysseus faced while far from home: "When it was the third watch of the night, and the stars had turned their course, Zeus, the cloud-gatherer, roused against us a fierce wind with a wondrous tempest, and hid with clouds the land and sea alike, and night rushed down from heaven."[14]

Vigilant sailors hoped to avoid these hazards by watching the sun and moon. They believed that these portended wind, rain, or storm when surrounded by halos or shrouded by hues of increasing darkness. Also, the ancient adage—"red sky at morning, sailors take warning"—expressed the common lore observed on land and sea. To this, Theophrastus added that if clouds obscured the sun as it set in the west, a storm was sure to come.[15]

Navigators also fixed their gaze on certain stars and constellations. If the twin stars called the Donkeys (Aselli) darkened or disappeared behind clouds, then rain or storm seemed imminent. Or, if clouds hid one of the Donkeys, then wind was expected to blow from the direction of the clear star toward the one obscured.

When sailing in southern waters near the Egyptian shore, a seaman could see the southern constellation known as Incense Altar (Ara). If this constellation shone bright, but banked with clouds, the crew prepared for a stormy night. Aratus urged sailors to heed this sign—to secure the rigging and stow the gear, and ready the ship for a brutal blast. If not, he warned, the tempest would "throw in turmoil all the sails," and the crew might "make their voyage all beneath the waves."[16]

In Aegean waters closer to home, if clouds encircled the loftier peaks, like Olympus, Athos, or Pelion, then wet and windy weather lay ahead. Or if clouds built above the mountains, or moved toward a ship at sea with frequent and frightening thunder and lightning, mariners prepared to head for harbor or hold on tight. But if the lightning proved less severe, with flashes at longer intervals, then the storm might arrive in a milder manner.[17]

On open waters, when a steady breeze suddenly ceased and left a ship carelessly rocking on calm seas, sailors watched for ripples on the waves that warned of a rush of wind from a new direction. Or if a breeze bore large amounts of thistle-down from a distant shore, the wind would soon blow with a violent blast. Furthermore, if wind arrived from the south or southeast, it would likely bring moisture and wet weather.[18]

Cranes assisted weather predictions just as well as they announced a change of seasons. Mariners knew that they only flew across the waters when weather would hold for a few more days. So, when the cranes' v-shape flight came into view, in fine formation, sailors knew that it meant mild conditions for the foreseeable future.[19]

But if the cranes turned about and headed back to land, or wheeled in flight with confused cries, a strong storm was brewing. Seabirds showed the same behavior. When they rushed to shore, fleeing in full retreat, a terrible tempest was on its way. Or, when dolphins often leapt and dove near land, rain or storm was close at hand.[20]

Mariners relied on weather lore and navigational knowledge handed down from captain to crew. They shared this lore, while gathering more at every Mediterranean port. In their heads they carried maps of rugged coastlines, lofty mountains, sandy shores, and dangerous shoals.

At sea, they kept a constant sense of their surroundings. They searched for telltale signs of changing currents and watched for ripples that might reveal an approaching gust or gale. They recognized the seasonal winds and detected slight shifts of direction, speed, and temperature.[21] With eyes squinted against the salty breeze and the glint of sunlight on the waves, they watched from heaving bows for familiar coastal features. Or they stood at the stern while sailing away from a shoreline, trying to align two land objects to direct a forward course.

In shallow waters, a captain commanded his crew to lower a bell-shape sounding weight, to strike the bottom and determine the depth. When drawn on deck, the tallow-filled bell revealed the local sand and shell. From this, the captain could conjecture the coastal location.

Many sailors preferred to remain as close to the coast as possible. This allowed them to follow familiar landmarks, or beat a hasty retreat from the angry sea. The Mediterranean offered many sandy shores for beaching boats and escaping storms. It also displayed a wide array of notable natural features—from high headlands and mountain peaks to pleasant islands and points of land that jut into the sea. Some of these provided unique and colorful rocks and soils, as well as singular shapes.

Man-made landmarks also added to the allure of the coastline. On many headlands, temples, towers, and tombs loomed above. At Troy, after Achilles fell in battle, Agamemnon described how the Greeks burned the body and buried the bones, then "heaped up a huge and flawless tomb . . . on a projecting headland by the broad Hellespont—that it might be seen from far over the sea, both by men that now are and that shall be born hereafter." [22]

Coastal landmarks aided navigation. But cruising close to the coast could also bring countless dangers. At night or in a blinding fog, anxious crews tried to avoid running ashore by lowering a sounding weight to determine diminishing depths of water.

Sometimes sailors, lost in the dark or fog near land, found familiar patterns of waves or currents, or coastal breezes that suggested the safest route. They also discerned the location of land by the lighter hues of shallow water; or the sound of waves breaking on the beach; or the smell of fields, forests, flocks, or fires. With a torch tied to the bow to light the way, they inched fearfully forward through the gloom, longing for the light of day.

In addition to the dangers of darkness and fog, coastlines suffered the wrath of stronger, shifting winds and currents, and larger waves that slapped the shores. Storms also formed where earth and water met, due to atmospheric contrasts of hot and cold over land and sea. As if all this was not enough, pirates infested hidden coves and harbors along the coast.

For all these reasons, seasoned seamen often preferred to sail in open waters. Here they followed the steady Etesian Winds of summer to full advantage. Strabo said that in summer, "the sea routes all pass through a zone of fair weather, particularly if the sailor keeps to the high seas. . . . And further, the winds on the high seas are regular."[23]

Many mariners also took to open waters in the face of approaching storms. With no time to run ashore or into harbor, they rushed their ships into surging seas to avoid being beached or wrecked upon the rocks. In this situation, a seaman who dared to leave land behind, and brave the heavy winds and waves, must have had a cool head indeed!

Some sailors felt at home on open seas, out of sight of land, and learned to rely on more than coastal landmarks. They discerned locations by close observations of local currents. Or they noted favorite feeding grounds where marine animals and seabirds assembled.

Furthermore, the flights of seabirds to and from these sites gave navigators a clue to the closest coastline. In desperate situations, ships sometimes followed a flock, or even the route of a single bird, to reach land. Some crews carried caged birds to release for the same purpose. Once the bird gained altitude and spied a place to roost, it flew there straight away.

Other signs of land appeared in the clear blue skies of summer. As moist sea breezes blew ashore, and warm and cool air masses mingled, clouds began to build. Before long, they billowed to lofty elevations above the land and became beacons for ships far out to sea. In addition, smoke sometimes streaked across the sky to provide a more certain sign of land.

Mariners carefully considered these signs, and also devised their own aids for navigation. The periplus (pl. peripli) served as a written record of distances and directions from port to port, or from one landmark to another over open water. These peripli noted the distance as numbers of days or stadia sailed.

The stadion (pl. stadia) was a unit of measure, typically of some 183 meters.[24]

Peripli provided nothing more than loose approximations, due to uncertainties involved in determining distance and direction traveled. These uncertainties came from constant variations in speed and direction of winds and currents, or cargo weight, or ship design, or a host of other factors known or unknown at the time. On top of this, most measurements amounted to mere guesses. No instrument had yet been made to mark them.

The best guess of distance came from sailing past some floating object to ascertain the ship's speed. This, multiplied by estimated time at sea provided the total distance. No solid measurement appeared in the whole equation. The geographer Strabo grumbled, rightly so, at the maddening lack of precision, saying, "Every writer does not agree with every other, particularly about the distances, as I often say."[25] Even so, peripli offered the best existing estimates of the time.

Sailors seemed to respond to the call of the sea without losing sleep over these navigational vagaries. The same courageous souls who chose to sail in open waters, far from land, also chose to navigate the seas at night. When seamen such as these made landfall, they did so not for fear of the dark, but to rest and feed the weary crew or to resupply the ship. They knew that the best way to get from point A to B was to ply the open sea day and night.

Vessels that ventured into the darkness often chose calm, clear, summer skies, far from coastal dangers. Most were merchant ships motivated by trade. Some were ships of war slipping stealthily toward an unsuspecting enemy. Nighttime navigation actually became quite common and dated back at least to the time of Homer.

At night, navigators gained the advantage of seeing the shining stars and constellations. These offered the best celestial compass. The Big and Little Bears denoted the cardinal direction north. South simply lay the opposite way. The rising and setting zodiacal constellations marked east and west.

During the day, the sun served the same purpose—rising in the east and setting in the west. At noon, its highest point in the sky marked the cardinal direction south. Some sailors, while resting during days spent ashore, used the gnomon—a simple stick stuck straight into the sand. With this, they noted the time of noon by the shortest length of the sun's shadow, and, at the same time, determined the direction south.[26]

When Odysseus was lost at sea, and longed to return to his island home of Ithaca, he used the stars and relied on the big Bear (Ursa Major) to mark the north at night while he sailed swiftly toward the east. Homer noted how "noble Odysseus spread his sail to the breeze; and he sat and guided his raft skillfully with the steering oar; nor did sleep fall upon his eyelids, as he watched the Pleiades, and late-setting Bootes, and the Bear.... For this star, Calypso, the beautiful goddess, had bidden him to keep on the left hand as he sailed over the sea."[27]

Homer's description suggests an event that occurred at summer solstice—at the start of safe sailing season. At that time of the year, with Odysseus sailing eastward and the Bear to his left, he would have seen the Pleiades ascend above the eastern horizon a few hours ahead of the sun. Arcturus, the brightest star of Bootes, would have set in the northwest an hour later—accurately matching Homer's description.

In later centuries, Phoenician navigators denoted the direction north, with greater precision, by relying on the Little Bear (Ursa Minor). Thales—the famed astronomer-philosopher—was first to reveal to his fellow Greeks that the sailors of Phoenicia followed this constellation as a more accurate guide in the sky.[28] Eratosthenes agreed that around this constellation "the entire universe appears to revolve," because it effectively marks the celestial North Pole.[29]

Aratus explained:

> It is by [the big Bear] that the Achaeans [Greeks] on the sea divine which way to steer their ships, but in [the Little Bear] the Phoenicians put their trust when they cross the sea. But [the big Bear], appearing large at earliest night is bright and easy to mark; but the other is small, yet better for sailors: for in a smaller orbit wheel all her stars. By her guidance, then, the [Phoenician] men of Sidon steer the straightest course.[30]

Thales, Eratosthenes, and Aratus correctly noted the more northerly abode of the Little Bear. Today, the tip of her tail is marked by Polaris—the North Star.[31]

Some sailors used constellations for more than finding cardinal directions. They knew how to sight the sun or stars to estimate the ship's present latitude—its position north to south. If they determined that their latitude matched that of a port of call, they could continue cruising along that line, east or west, to intercept the city. Seamen knew the latitudes of several port cities, where citizens used a gnomon—a simple vertical stick—to mark the sun's angle on certain days of the year. For example, when the sun cast a noon shadow on summer solstice, the angle served as a measure of latitude.

By similar use of a gnomon, the Greek mathematician Eratosthenes (276–194 BC) measured the angle of the sun's shadow and used that angle to calculate the Earth's circumference. Centuries later, scientists learned that his answer proved accurate to 99.2 percent—a phenomenal feat.

Eratosthenes based his calculation on the theory of a spherical Earth—a theory that dated back at least three centuries earlier, to

Pythagoras. Eratosthenes also operated on the valid premise that the sun is at such an extreme distance that it sends virtually parallel rays to the Earth's surface. When he heard reports of sunlight reaching the bottom of deep wells at Syene (present Aswan) at noon on summer solstice, he realized that the sun stood straight above on that day, at that latitude. But farther north in Alexandria, where he lived, the sun at noon on summer solstice cast a gnomon shadow at an angle of one-fiftieth of a circle (7.2 degrees). From this, he rightly determined that the north–south distance from Alexandria to Syene equaled one-fiftieth of the Earth's entire circumference.

Based on a camel trip between these two cities, Eratosthenes surmised a distance of 5040 Egyptian stadia (one Egyptian stadion equaled 157.5 meters). Fifty times 5040 stadia gave the circumference of the Earth at 252,000 stadia, or 39,690 kilometers. The north–south circumference of the Earth is actually 40,008 kilometers. His calculation was off by less than one percent![32]

Mariners, who estimated latitude while on their ships at sea, did so much more casually, with a far greater margin of error. With a hand held out at arms-length, they counted finger-widths to note the sun's noontime height above the horizon. At night, they marked the Little Bear in the same manner.

Latitude helped sailors guess the ship's position. But they also needed to determine time at sea, and the distance traveled. They gauged the passage of time by watching the sun climb in the east and descend in the west. At night, the constellations and stars of the zodiac pursued the same path as the sun and provided a similar measure of time.

Other constellations assisted in the same way. Aratus declared how readily "can the sailor on the open sea mark the first bend of the River [the constellation Eridanus] rising from the deep, as he watches for Orion himself to see if he might give him any hint of the measure of the night or of his voyage."[33] This ancient concept of marking a beginning, or base time, and a present time, to determine distance and present location, helped astronomers two millennia later to finally calculate longitude.

Another means of measuring time involved watching the Little Bear as it slowly circled the northern sky. From this, a navigator could tell "when it was the third watch of the night, and the stars had turned their course" around the celestial North Pole.[34] Now they knew how much of the night remained before rosy-fingered dawn would arrive to disperse the darkness.

With all these skills at their disposal, ancient sailors determined the proper seasons and weather conditions for sailing. They noted locations at sea by watching coastal landmarks, taking seafloor soundings, or observing currents and clouds. They surmised cardinal directions by means of prevailing winds and annual bird migrations. They measured movements of the sun, moon, and stars to discern direction, latitude, and the passage of time. The celestial lights, both day and night, gave mariners the gift of orientation in space and time, and allowed them to answer the vital navigational questions of where and when.

ACKNOWLEDGMENTS

My sincere appreciation to the gracious, gift-giving people of Greece, who assisted and befriended my wife and me during our travels and research. In particular, I wish to thank those who live in and near Athens, Piraeus, Eleusis, Laurium, Kamariza, Olympia, Arcadia, Epidaurus, Mycenae, Delphi, Thebes, Mt. Helicon, Plataea, Leuctra, Marathon, Oropus, Thermopylae, Mt. Olympus, Volos, Mt. Pelion, Chorefto, and the lovely islands of Crete, Santorini, and Ithaca.

A special thanks to Anthony B. Kaye, PhD, professor of physics and astronomy at Texas Tech University, who gave generously of his time to explore, in depth, the various methodologies used to determine the ancient dates of solstices, equinoxes, and apparent star risings and settings. From this, he defined the variables involved and developed an accurate means of calculating these celestial events in relation to our modern calendar dates.

Above all, I am grateful for the love and support of my wife and best friend, Vicki, in whose memory this book is dedicated. Always patient and undaunted, she remained my constant companion across the Mediterranean and around the world in pursuit of research and adventure.

APPENDIX 1: GREEK CONSTELLATION, ASTERISM, AND STAR NAMES

TRANSLATION FROM GREEK ASTERISM/STAR	GREEK NAME	LATIN/MODERN NAME
Andromeda	Ανδρομεδα	Andromeda
Antinous	Αντινοοσ	Antinous
Archer	Τοξοτησ	Sagittarius
Ares Counterpart	Ανταρησ	Antares
Argo	Αργω	Argo Navis
Arrow	Οιστοσ	Sagitta
Bear	Αρκτοσ	Ursa Major
Bear Guard	Αρκτουροσ	Arcturus
Bird	Ορνισ	Cygnus
Bootes	Βοωτησ	Bootes
Bull	Ταυροσ	Taurus
Canopus	Κανωβοσ	Canopus
Cassiopeia	Κασσιεπεια	Cassiopeia
Centaur	Κενταυροσ	Centaurus
Cepheus	Κηφευσ	Cepheus
Charioteer	Ἡνιοχοσ	Auriga
Claws	Χηλαι	Libra
Crab	Καρκινοσ	Cancer
Crater	Κρατηρ	Crater
Crow	Κοραξ	Corvus
Dog	Κυων	Canis Major
Dog Star	Κυων	Sirius
Dolphin	Δελφιν	Delphinus
Donkeys	Ονοι	Aselli
Dragon	Δρακων	Draco
Eagle	Αετοσ	Altair
Eagle	Αετοσ	Aquila
Ear of Grain	Σταχυσ	Spica
Fishes	Ιχθυεσ	Pisces
Goat	Αιξ	Capella
Goat Horn	Αιγοκερωσ	Capricornus

TRANSLATION FROM GREEK ASTERISM/STAR	GREEK NAME	LATIN/MODERN NAME
Gorgon	Γοργω	Algol
Hare	Λαγωοσ	Lepus
Herald of the Dog	Προκυων	Canis Minor
Herald of the Dog	Προκυων	Procyon
Herald of the Vintage	Προτρυγητηρ	Vindemiatrix
Horse	Ἱπποσ	Pegasus
Horse Head	Ἱπποσ Προτομησ	Equuleus
Hyades	Ὑαδεσ	Hyades
Hydra	Ὑδρα	Hydra
Incense Altar	Θυμιατηριον	Ara
Kids	Εριφοι	Haedi
Kneeler	Εν γονασιν	Hercules
Lion	Λεων	Leo
Little Bear	Αρκτοσ μικρα	Ursa Minor
Lyre	Λυρα	Lyra
Lyre	Λυρα	Vega
Maiden	Παρθενοσ	Virgo
Manger	Φατνη	Praesepe
Northern Wreath	Στεφανοσ βορειοσ	Corona Borealis
Orion	Ωριων	Orion
Perseus	Περσευσ	Perseus
Pleiades	Πληιαδεσ	Pleiades
Prince	Βασιλισκοσ	Regulus
Ram	Κριοσ	Aries
River	Ποταμοσ	Eridanus
Scorpion	Σκορπιοσ	Scorpius
Sea Monster	Κητοσ	Cetus
Snake	Οφισ	Serpens
Snake Holder	Οφιουχοσ	Ophiuchus
Southern Fish	Ιχθυσ νοτιοσ	Piscis Austrinus
Southern Wreath	Στεφανοσ νοτιοσ	Corona Australis
Triangle	Τριγωνον	Triangulum
Twins	Διδυμοι	Gemini
Water Bearer	Ὑδροχοοσ	Aquarius
Wild Animal	Θηριον	Lupus

APPENDIX 2: MODERN CONSTELLATION NAMES

LATIN/ MODERN NAME	TRANSLATION FROM LATIN	TRANSLATION FROM GREEK
Andromeda	Andromeda	Andromeda
Antlia	Pump	
Apus	Bird of Paradise	
Aquarius	Water Bearer	Water Bearer
Aquila	Eagle	Eagle
Ara	Altar	Incense Altar
Argo Navis*	The Ship Argo	Argo
Aries	Ram	Ram
Auriga	Charioteer	Charioteer
Bootes	Bootes	Bootes
Caelum	Chisel	
Camelopardalis	Giraffe	
Cancer	Crab	Crab
Canes Venatici	Hunting Dogs	
Canis Major	Big Dog	Dog
Canis Minor	Little Dog	Herald of the Dog
Capricornus	Goat Horn	Goat Horn
Carina	Ship's Keel	
Cassiopeia	Cassiopeia	Cassiopeia
Centaurus	Centaur	Centaur
Cepheus	Cepheus	Cepheus
Cetus	Sea Monster	Sea Monster
Chamaeleon	Chameleon	
Circinus	Compass for Drawing	
Columba	Dove	
Coma Berenices	Berenice's Hair	
Corona Australis	Southern Wreath	Southern Wreath
Corona Borealis	Northern Wreath	Northern Wreath

LATIN/ MODERN NAME	TRANSLATION FROM LATIN	TRANSLATION FROM GREEK
Corvus	Crow	Crow
Crater	Wine Mixing Bowl	Crater
Crux	Cross	
Cygnus	Swan	Bird
Delphinus	Dolphin	Dolphin
Dorado	Swordfish	
Draco	Dragon	Dragon
Equuleus	Little Horse	Horse Head
Eridanus	Eridanus	River
Fornax	Furnace	
Gemini	Twins	Twins
Grus	Crane	
Hercules	Hercules	Kneeler
Horologium	Clock	
Hydra	Hydra	Hydra
Hydrus	Water Snake	
Indus	Indian	
Lacerta	Lizard	
Leo	Lion	Lion
Leo Minor	Little Lion	
Lepus	Hare	Hare
Libra	Scales	Claws
Lupus	Wolf	Wild Animal
Lynx	Lynx	
Lyra	Lyre	Lyre
Mensa	Table	
Microscopium	Microscope	
Monoceros	Unicorn	
Musca	Fly	
Norma	Level	
Octans	Octant	
Ophiuchus	Snake Holder	Snake Holder
Orion	Orion	Orion
Pavo	Peacock	
Pegasus	Pegasus	Horse

LATIN/ MODERN NAME	TRANSLATION FROM LATIN	TRANSLATION FROM GREEK
Perseus	Perseus	Perseus
Phoenix	Phoenix	
Pictor	Painter's Easel	
Pisces	Fishes	Fishes
Piscis Austrinus	Southern Fish	Southern Fish
Puppis	Ship's Stern	
Pyxis	Ship's Compass	
Reticulum	Net	
Sagitta	Arrow	Arrow
Sagittarius	Archer	Archer
Scorpius	Scorpion	Scorpion
Sculptor	Sculptor	
Scutum	Shield	
Serpens	Snake	Snake
Sextans	Sextant	
Taurus	Bull	Bull
Telescopium	Telescope	
Triangulum	Triangle	Triangle
Triangulum Australe	Southern Triangle	
Tucana	Toucan	
Ursa Major	Big Bear	Bear
Ursa Minor	Little Bear	Little Bear
Vela	Ship's Sail	
Virgo	Virgin	Maiden
Volans	Flying Fish	
Vulpecula	Little Fox	

*Argo Navis is not one of the eighty-eight modern constellations, but it numbers among the forty-eight classical Greek constellations. In the eighteenth century, it was divided into the modern constellations of Vela, Carina, Pyxis, Puppis, and the star, Eta Columbae.

APPENDIX 3: ANNUAL CELESTIAL EVENTS, NOTED BY HESIOD WITH ADDITIONS BY EUDOXUS[1]

DATE	CELESTIAL OBJECT	EVENT
Feb 25	Bear Guard	apparent achronycal rising marks grapevine pruning; approach of spring;[2] stormy season[3]
Mar 28	Ram	vernal equinox marks equal day and night; sun aligns with Ram[4]
Apr 5	Pleiades	heliacal setting marks about forty days until Pleiades heliacal rising at grain harvest[5]
May 16	Pleiades	heliacal rising marks grain harvest;[6] approach of summer[7]
June 29	Orion	heliacal rising marks grain winnowing[8]
June 30	Crab	summer solstice marks fifty days of fair sailing;[9] sun aligns with Crab[10]
July 20	Herald of the Dog	heliacal rising marks approach of Dog Days
July 28	Dog star	heliacal rising marks Dog Days of summer[11]
July 30	Lion	sun entering the Lion marks summer heat; forty-day Etesian Winds; tricky sailing[12]
Sept 13	Herald of the Vintage	heliacal rising marks grape harvest[13]
Sept 21	Bear Guard	heliacal rising marks grape harvest;[14] approach of stormy season[15]
Sep 30	Claws	autumnal equinox; sun aligns with Claws[16]
Nov 3	Pleiades	apparent cosmical setting marks plowing; end of sailing season;[17] approach of winter[18]
Nov 11	Hyades	apparent cosmical setting marks plowing[19]
Nov 25	Orion	apparent cosmical setting marks plowing[20]
Dec 13	Eagle	heliacal rising marks stormy weather[21]
Dec 27	Goat Horn	winter solstice marks sixty wintry days;[22] sun aligns with Goat Horn; short days; bitter cold; dangerous sailing[23]

APPENDIX 4:
THE HARMONIOUS WHOLE

The holistic theory of the cosmos, advanced by Thales of Miletus (c. 624–c. 546 BC) and Anaximenes of Miletus (c. 585–c. 528 BC), became a basic tenet of Greek philosophy (see Introduction). The following writings indicate the same:

Anaximander of Miletus (c. 610–c. 546 BC):

> *The Infinite (apeiron) . . . eternal and ageless . . . is the cause (arche) and first element (stoicheion) of things . . . from which arise all the heavens and the worlds within them. . . . And into that from which things take their rise they pass away once more.*
>
> [Anaximander, *Fragments*, in *Early Greek Philosophy*, quoted by Theophrastus and Simplicius, translated by John Burnet (London: Adam and Charles Black, 1930), p. 52.]

Pherecydes of Syros (c. 580–c. 520 BC):

> Divine Love created the cosmos. *Putting together the cosmos from opposites, he brought it into harmony and love, and sowed likeness in all things and oneness pervading the whole.*
>
> [Pherecydes, *Fragments*, in *Pherekydes of Syros*, quoted by Proclus, translated by Hermann Schibli (Oxford: Clarendon Press, 1990), p. 168.]

Xenophanes of Colophon (c. 570–c. 475 BC):

> *There is one god (theos), among gods and men the greatest, not at all like mortals in body or in mind. He sees as a whole, thinks as a whole, and hears as a whole.*
>
> [Xenophanes, *Fragments*, in *Ancilla to the Pre-Socratic Philosophers*, quoted by Clement and Sextus Empiricus, translated by Kathleen

Freeman (Cambridge: Harvard University Press, 1962), p. 23; Hermann Diels, *Die Fragmente der Vorsokratiker* (Berlin: Weidmann, 1964), p. 135.]

Heraclitus of Ephesus (c. 535–c. 475 BC):

When you have listened, not to me but to the Logos (ultimate reason), *it is wise to agree that all things are one.*

Heraclitus and other Presocratic philosophers further insisted that things in seeming opposition actually intertwine as one. One of Heraclitus' analogies stated:

The way up and down is one and the same.

[Heraclitus, *Fragments*, in *Ancilla to the Pre-Socratic Philosophers*, quoted by Hippolytus, translated by Kathleen Freeman (Cambridge: Harvard University Press, 1962), pp. 28–29; Diels, *Die Fragmente der Vorsokratiker*, pp. 161, 164.]

Universal oneness remained a common theme of the Presocratic Philosophers. They used various descriptors for the *infinite, indefinable One*. These included Hesiod's *Chaos* (meaning *undefined* or *formless*), Anaximander's *Infinite* (apeiron), Pherecydes' *Divine Love* (eros), Xenophanes' *God* (theos), Heraclitus' *Ultimate Reason* (logos), Parmenides' *Being* (einai), and Anaxagoras' *Mind* (nous). These philosophers, along with Melissus, Empedocles, Philolaus, and Diogenes, spoke often of the *One* and the *Whole*.

The three Athenian philosophers—Socrates (c. 469–399 BC), Plato (c. 424–c. 348 BC), and Aristotle (384–322 BC)—agreed with their predecessors that there existed one eternal, primal source, which was an intangible and indescribable spirit. Plato called this spirit *agathos*—the *good*. He provided a fine example of this aspect of Presocratic and Socratic thought in his *Allegory of the Cave*. [Plato, *The Republic*, translated by Paul Shorey (Cambridge: Harvard University Press, 1970), sections 514–518.]

Even though this spirit was indescribable in human terms, the Athenian philosophers asserted that awareness of it remained within reach of persons possessing exceptional insight. For them, it was possible to transcend the superficialities and anxieties of the world, in their lifetime, and gain a glimpse of the *good*. Further, they believed the *good* would become fully evident in the eternal life that followed.

Socrates said that those who keep their minds fixed on the *good* of the spiritual realm need not fear—for no lasting "evil can happen, either in life or in death." He believed that the world was transitory and the human body perishable, but that the spirit within was immortal and bound for a higher existence. Consequently, he pleaded with his grieving students, at the time of his execution, to:

> *Be of good cheer . . . and say that you are burying my body only.*
>
> [Plato, *Phaedo*, in *The Dialogues of Plato*, translated by B. Jowett (New York: Random House, 1937), section 116. Socrates left no written records, but Plato often quoted him, and provided insight into the mind of his famous mentor.]

Socrates' student—Plato—expanded upon his teachings by stressing the spiritual nature of reality. The world, he said, is but a dim, deficient shadow; so the worthiest human endeavor is to seek the spiritual realm. Humans inherently yearned for this, but to reach it required transcending worldly distractions. One needed to shun the pursuit of power, possessions, and selfish strife. Instead, one should live simply, in peace, while seeking the spiritual *good*. Plato wrote:

> *My dream, as it appears to me, is that . . . the idea of good (agathos) . . . is indeed the cause for all things—of all that is right and beautiful, giving birth in the visible world to light and the author of light [the sun]; And itself in the intelligible world being the authentic source of truth and reason; And that anyone who is to act wisely in private or public must have caught sight of this. . . . That those who have*

attained to this height are not willing to occupy themselves with the affairs of men, but their souls ever feel the upward urge and the yearning for that sojourn above.

[Plato, *Republic*, section 517.]

When a man is always occupied with the cravings of desire and ambition, and is eagerly striving to satisfy them, all his thoughts must be mortal.... But he who has been earnest in the love of knowledge and of true wisdom ... since he is ever cherishing the divine power, and has the divinity within him in perfect order, he will be perfectly happy.

[Plato, *Timaeus*, section 90.]

Plato's student—Aristotle—added that the physical world should not be neglected in favor of the spiritual realm. After all, the divine spirit is tangible in the natural world just as it is intangible in the human mind and soul. The physical and spiritual universe are one harmonious whole. Aristotle said:

God (theos) is *the first principle, upon which depends the sensible [sensory] universe and the world of nature.... We hold then that God is a living being, eternal, most good; and therefore life and a continuous eternal existence belong to God; for that is what God is.*

[Aristotle, *Metaphysics*, translated by Hugh Tredennick (Cambridge: Harvard University Press, 1947), section 12.7.7–9.]

This *supreme good* encompasses the cosmos and is also the *orderly arrangement of its parts....* Thus, *everything is ordered together to one end,* and *everything contributes to the good of the whole.*

[Aristotle, *Metaphysics*, 12.10.1–4.]

APPENDIX 5: MAPS

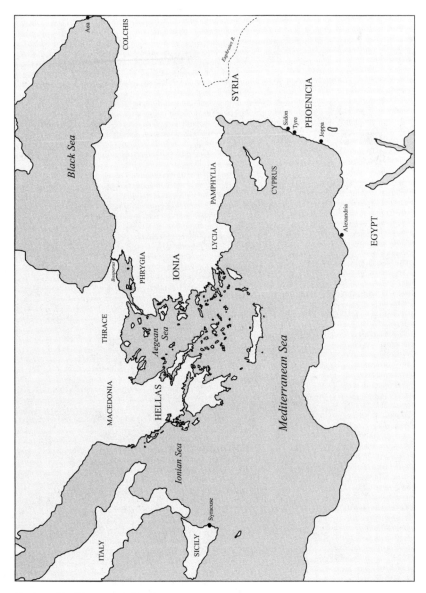

Eastern Mediterranean Sea

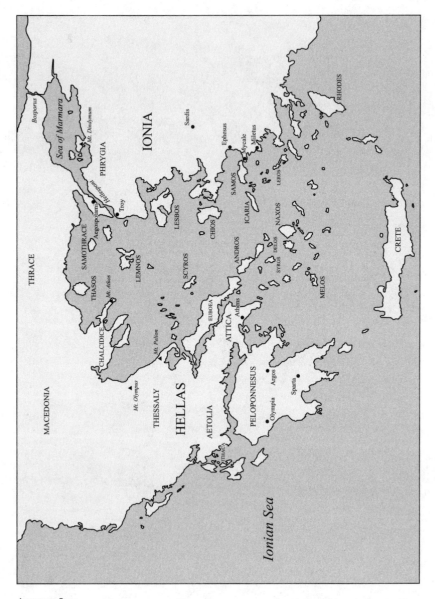

Aegean Sea

APPENDIX 5

NOTES

INTRODUCTION

1. Manilius, *Astronomica*, translated by G. P. Goold (Cambridge: Harvard University Press, 1977), 1.717.

2. Aratus, *Phaenomena*, in *Callimachus; Lycophron; Aratus*, translated by G. R. Mair (Cambridge: Harvard University Press, 1955), 373.

3. [Homer, *Iliad*, translated by A. T. Murray and William F. Wyatt (Cambridge: Harvard University Press, 1999), 18.483]; some evidence suggests that Paleolithic peoples defined Ursa Major, Orion, and the Pleiades before the last Ice Age migrations from Asia to America. The universal recognition of these star groups as a bear, a hunter, and seven sisters lends credibility to the theory. [William Gibbon, "Asiatic Parallels in North American Star Lore: Ursa Major," *Journal of American Folklore*, 77 no. 305 (1964): 236–250; William Gibbon, "Asiatic Parallels in North American Star Lore: Milky Way, Pleiades, Orion," *Journal of American Folklore*, 85 no. 337 (1972): 236–247; Maud Makemson, "Astronomy in Primitive Religion," *Journal of Bible and Religion*, 22 no. 3 (1954): 163–171]; evidence also exists of possible constellation diagrams depicted in Ice Age cave paintings in France and Spain, most notably on the walls of Lascaux Cave. One aurochs (bull no. 18) at Lascaux closely parallels the orientation of the constellation Taurus, with its horns lowered and face speckled with a pattern akin to the asterism named Hyades. Above its back is another pattern resembling the six visible stars of the Pleiades—an asterism found on the hump of the constellation Taurus. [Michael Rappenglueck, "The Pleiades in the 'Salle des Taureaux,' Grotte de Lascaux," www.infis.org/research.]

4. An asterism is a named group of stars not designated as a constellation. The Pleiades and Hyades, both within the constellation Taurus, serve as examples.

5. Aratus, *Phaenomena*, 531.

6. Ibid, 452.

7. Ibid, 454.

8. Our word *planet* comes from the Greek verb *planasthai*, which means *to wander*.

9. [Homer, *Iliad*, 23.226]; Pindar described Phosphoros as *splendid to behold among the other stars*. [Pindar, *Isthmian Odes*, in *Nemean Odes; Isthmian Odes; Fragments*, translated by William Race (Cambridge: Harvard University Press, 1997), 4.24.]

10. Homer, *Odyssey*, translated by A. T. Murray and George E. Dimock (Cambridge: Harvard University Press, 1995), 13.93.

11. Homer, *Iliad*, 22.316.

12. Pythagoras was the first to identify the five wandering planets and to hypothesize that the Earth was a sphere. Venus' mysterious changing position in the sky, from morning to evening, was later explained by the planet's close orbit around the sun.

13. Aphrodite, Zeus, Ares, and Hermes represented four of the fourteen major Greek gods. Apollo, the sun god, and his twin sister Artemis, the moon goddess, represented two others. Among the modern planets, Uranus (Father Sky) was the husband of Gaea (Mother Earth). They were the parents of the Titans and grandparents of the gods. Poseidon (Neptune) ruled as god of the sea and Pluto as god of the underworld. The gods that remained unrepresented among the planets included Athena—goddess of wisdom and skill-at-arms; Demeter—goddess of farming and harvest; Hestia—goddess of hearth and home; Hera—goddess of women and childbirth; Hephaestus—god of fire and metallurgy; and Dionysus—god of wine and revelry.

14. See the *Oxford Classical Greek Dictionary* and *Oxford English Dictionary*.

15. Plato, *Timaeus*, in *The Dialogues of Plato*, translated by B. Jowett (New York: Random House, 1937), section 40.

16. Ptolemy catalogued many stars that depicted animal anatomy, specifically denoting the head, skull, hair, forehead, horn, tip of the horn, temple, eye, eyebrow, ear, snout, nostril, cheek, mouth, muzzle, bird's beak, tongue, jaw, neck, bend of the neck, horse's mane, fish gills, body, chest, bird's breast, heart, belly, navel, back, upper back (between the shoulders), shoulder blade, dorsal fin, flank, hindquarters, rump, buttocks, forelegs, hind legs, armpit, thigh, back of the thigh, knee, knee bend, hock, wing, upper wing, bend of the wing, under the wing, wing tip, wing feathers, hoof, paw, claw, tail, base of the tail, bend of the tail, tip of the tail, tail fin, and scorpion's stinger. [Ptolemy, *Almagest*, in *Syntaxis Mathematica*, translated by J. L. Heiberg (Leipzig: B. G. Teubneri, 1898); Ptolemy, *Almagest*, in *Ptolemy's Almagest*, translated by G. J. Toomer (Princeton: Princeton University Press, 1998).]

17. Ptolemaic stars showing anthropomorphic anatomy depicted the head, skull, face, eye, neck, body, chest, female breast, back, upper back (between the shoulders), shoulder blade, belly, groin, side, shoulder, arm, armpit, elbow, upper arm, forearm, wrist, hand, buttocks, hip, legs, thigh, back of the thigh, knee, knee bend, lower leg, calf muscle, shin, ankle, foot, heel, and arch. Clothing is sparsely distinguished and includes an apron (or tunic), girdle, garment's lower hem, coat, cloak with tie strings, belt, and tiara. Accoutrements are more abundant and include a shepherd's staff, club, sword (with hilt), dagger, shield, bow (with grip), arrow (with arrowhead, shaft, and notch), animal pelt (used as a garment or arm-guard), fish-

ing line knotted together, cup or bowl (with base, rim, and handles), lyre (with tortoise-shell body, cattle-horn neck, and bridge), incense burner (with base, burning pan, and *burning apparatus*—perhaps an incense or wick holder), thyrsus (a branch adorned with vine leaves and pine cones carried by followers of Dionysus), throne, and ship (with deck, mast, mast-holder, poop-deck, stern-post, keel, and steering oars). Natural objects are limited to water sources and include a flow of water and the bend of a river in the constellations Aquarius and Eridanus.

18. Every Greek, from early childhood, had heard the tales of Heracles (Hercules). Now they could see their hero in the northern sky at night, wreathed in a robe of heavenly light. Beneath him, a slithering serpent, named Draco, appeared as a coiling stream of stars. With a little imagination, the dramatic story of old that told of the deadly duel between the two now came to life. With ease, one could see Heracles kneeling to pin the beast, while raising a club to strike it down. The constellations of Orion and Scorpius stand as another example. The two are easy to see among the stars as Scorpius pursues the path of his eternal enemy: Orion. Thus, they provide a perfect portrayal of the oft-told tale of Orion's fateful plight.

19. James Evans, *The History and Practice of Ancient Astronomy* (Oxford: Oxford University Press, 1998), p. 39.

20. Aristotle called astronomy *that branch of mathematical science which is most akin to philosophy.* [Aristotle, *Metaphysics*, translated by Hugh Tredennick (Cambridge: Harvard University Press, 1947), section 12.8.7.]

21. See Appendix 4.

22. Plato's ideal person was well rounded and whole: an athlete, musician, and seeker of knowledge, wisdom, and truth. The ideal person knew how to stand silent and still in order to listen and learn; unlike the foolish man who runs his mouth non-stop—trying to make up for all he is not. Plato complained about those who *go ringing on in a long harangue, like brazen pots, which when they are struck continue to sound unless someone puts his hand upon them.* [Plato, *Protagoras*, in *The Dialogues of Plato*, translated by B. Jowett (New York: Random House, 1937), section 329]; The ideal person also had a broad perspective, based on a broad background that included hard physical labor, practical sense, theoretical knowledge, and spiritual yearning. Plato said, *He who is careful to fashion the body, should in turn impart to the soul its proper motions, and should cultivate music and all philosophy, if he would deserve to be called truly fair and truly good.* [Plato, *Timaeus*, section 88.]

23. Pausanias wrote, *In the fore-temple at Delphi are written maxims useful for the life of men, inscribed by those whom the Greeks say were sages. These were: from Ionia, Thales of Miletus [et al.]. . . These sages, then, came to Delphi and dedicated to Apollo the celebrated maxims, "Know thyself," and "Nothing in excess."* [Pausanias, *Description of Greece*, translated by W. H. S. Jones (Cambridge: Harvard University Press, 1955), 10.24.]

24. Archaeological evidence, from the Paleolithic to the present, often suggests a standard belief in a single all-encompassing deity, whether it be a Mother Earth goddess, a Father Sky god, or an undefined concept that becomes manifest in standardized rituals and burials. In addition, earliest encounters with prehistoric societies, like those in Africa and America, evince a global belief in a single, beneficent, all-embracing deity.

25. There is a logical explanation for the numerous stories of philandering gods. Archaic Greece was characterized by small city-states, each of which venerated its own deity. More often than not, these were local earth goddesses. As Zeus's reputation as the supreme god spread, he supplanted many of the local goddesses. The logical way to affect this change was to explain in a myth that Zeus had cohabitated with the local goddess—an act that would have established his paternal dominance. As the same story became repeated in different localities, Zeus gained a reputation as a philanderer.

26. Other Mediterranean cultures also denoted asterisms and constellations to mark cardinal directions. Richard Allen noted that: *the three constellations mentioned in the biblical Book of Job [Chapters 9:9 and 38:31–32; written c. 600 BC] and Book of Amos [Chapter 5:8; written c. 750 BC] fitly represent the cardinal points in the sky: the Bear in the north, Orion in the south, and the Pleiades rising and setting in the east and west.* [Richard Allen, *Star Names: Their Lore and Meaning* (New York: Dover Publications, 1963), p. 309.]

27. Aratus, *Phaenomena*, 5.

28. Homer, *Odyssey*, 12.310, 14.482.

29. Aratus, *Phaenomena*, 740.

30. [Aratus, *Phaenomena*, 559]; Due to the Earth's approximate 365-day orbit around the sun, each star and constellation appears on the horizon about four minutes (or one angular degree) earlier each night—that is, farther ahead of the sun—as it begins a westward journey across the sky.

31. In the Sumerian *Epic of Gilgamesh*—the oldest existing written story—a falling star (meteor) is a herald of good fortune. The *Epic* also mentions that stars offer protection as *Watchmen of the Night*. [*Epic of Gilgamesh*, translated by N. K. Sandars (New York: Penguin Books, 1972), Cuneiform Tablet 1, p. 66; Cuneiform Tablet 3, p. 75.]

32. https://commons.wikimedia.org/wiki/file:Atlante.JPG.

CHAPTER I

1. Homer said of the *great-hearted Erechtheus* that Athena *settled* him in her *own rich shrine*, suggesting some validity to the tradition that Erechtheus was buried at the temple on the Acropolis. [Homer, *Iliad*, 2.547]; In the *Odyssey*, Athena left Odysseus with King Alcinous and came to *broad-wayed Athens, and entered the well-built house of Erechtheus*—the temple he had dedicated. [Homer, *Odys-*

sey, 7.80]; A much later temple, begun in 421 BC, was named the Erechtheum after him and still stands on the Acropolis. Erechtheus is sometimes called Erichthonius.

2. The *Charioteer of Delphi* once stood in the Sanctuary of Apollo at Delphi.

3. Five of the Caryatids are now preserved in the Acropolis Museum while reproductions stand in their place at the Erechtheum. The sixth Caryatid stands alone in the British Museum near the Elgin Marbles. Carya is the Greek word for *nut*. Today the name Carya is applied to the genus of nut-bearing trees represented by hickories worldwide, including the North American pecan—Carya illinoinensis. The genus Corylus probably derives from Carya as well and is represented by Corylus avellana—the hazelnut tree common in Eurasia.

4. The manger is marked by the open star cluster now called Praesepe (Greek for *manger*).

5. Ptolemy referred to the larger fish beneath Pegasus as being in advance of the other. Presumably, this one is Aphrodite and the smaller one beneath Andromeda is Eros.

6. Piscis Austrinus is marked by the bright star Fomalhaut.

7. Apollonius Rhodius, *The Argonautica*, translated by William H. Race (Cambridge: Harvard University Press, 2008), 4.933.

8. Homer, *Iliad*, 20.233.

9. Ganymede is an asterism of the constellation Aquila. Ptolemy later referred to the asterism as Antinous.

10. The Greek term for incense altar, *thymiaterion*, likely comes from the use of thyme as a common incense. Thyme was thought to instill courage.

11. Homer, *Iliad*, 9.496.

CHAPTER 2

1. The complete passage states: *Those who do not turn aside from justice at all, their city blooms and the people in it flower. For them, Peace, the nurse of the young, is on the earth, and far-seeing Zeus never marks out painful war; nor does famine attend straight-judging men, nor calamity, but they share out in festivities the fruits of the labors they care for. For these the earth bears the means of life in abundance, and on the mountains the oak tree bears acorns on its surface, and bees in its center; their woolly sheep are weighed down by their fleeces; and their wives give birth to children who resemble their parents. They bloom with good things, continuously.* [Hesiod, *Works and Days*, in *Theogony; Works and Days; Testimonia*, translated by Glenn W. Most (Cambridge: Harvard University Press, 2006), 225]; The ancient Greeks later quoted this passage in praise of the goddess Dice. [Hesiod, *Testimonia*, in *Theogony; Works and Days; Testimonia*, translated by Glenn W. Most (Cambridge: Harvard University Press, 2006), 107.]

2. Astraea is often identified as Dice. The ancient Greeks and Romans knew Dice (Justice), Eunomia (Order), and Eirene (Peace) collectively as the Horae—the daughters of Zeus and Themis. Richard Allen noted that Virgo, or Dice, was also known as *Astraea, the starry daughter of Themis, the last of the celestials to leave the earth . . . when the Brazen Age began.* The name *Astraea* literally means *starry* in Greek. The constellation Virgo is the oldest existing *representation of innocence and virtue* in the human record. [Allen, *Star Names*, p. 422.]

3. Hesiod said, *Eos, a goddess bedded in love with a god, bore to Astraeus the strong-spirited winds, clear Zephyrus and swift-pathed Boreas and Notus; and after these the Early-born one (Eos) bore the star, Dawn-bringer (Phosphoros), and the shining stars with which the sky is crowned.* [Hesiod, *Theogony*, in *Theogony; Works and Days; Testimonia*, translated by Glenn W. Most (Cambridge: Harvard University Press, 2006), 378.]

4. Aratus, *Phaenomena*, 107.

5. In ancient Greece, Vindemiatrix's heliacal rising marked the beginning of the grape harvest on approximately September 13.

6. The nine Muses—daughters of Zeus and Mnemosyne (Memory)—included Calliope (Muse of epic poetry), Clio (history), Erato (lyric and love poetry), Euterpe (instrumental music, and inventor of the flute and woodwinds), Melpomene (tragic drama), Polyhymnia (singing and oratory), Terpsichore (dance), Thalia (comedic drama), and Urania (astronomy). These represent the enlightened accomplishments of humans that transcend the tangible and mundane. History and astronomy are aptly included, as they represent a broadened human perspective in time and space, and like the arts, open the mind to a more meaningful existence.

7. Pegasus, too, had danced in mirth and ecstasy in company with the Muses on Mount Helicon. During an especially robust prance, he pierced the ground, from which came a sparkling stream known thereafter as the Hippocrene. The name literally translates as *fountain of the horse*. The fountain served as a source of sustenance and poetic inspiration for the Muses. Hesiod said the Muses bathe in the sacred waters of the Hippocrene, then *perform choral dances on highest Helikon, beautiful, lovely ones . . . and dance on their soft feet around the violet-dark fountain* and the altar of Zeus. [Hesiod, *Theogony*, 1–8.]

8. Apollonius Rhodius, *Argonautica*, 1.553.

9. Aratus, *Phaenomena*, 130.

10. Homer, *Iliad*, 11.152.

11. Ibid, 1.3.

12. [Aesop, *Fables*, in *Babrius and Phaedrus*, compiled by Babrius, translated by Ben Edwin Perry (Cambridge: Harvard University Press, 1965), Fable 79: The Dog and the Shadow]; Aesop is a shadowy figure mentioned by Herodotus, Aristotle, Plutarch, and other reliable sources. These suggest that he was born about 620 BC in Thrace on the Black Sea, perhaps at Mesembria (present-day Nesebar, Bul-

garia). Herodotus and Aristotle said he was later a slave in Samos, but was freed. Plutarch added that he was executed at Delphi. By the fifth century BC, an Aesop legend developed on the island of Samos. His fables probably circulated as oral traditions before being written down. Over the ensuing centuries, the original writings were lost; however, later versions survived. The earliest surviving versions are by Babrius in Greek and Phaedrus in Latin, both of whom lived sometime during the first two centuries AD. The two probably worked independently from extant copies. Babrius appears as the more reliable source and claims only to make the fables more poetic, while Phaedrus inserts much new but inferior material.

13. Herodotus, *The Histories*, translated by Robin Waterfield (Oxford: Oxford University Press, 1998), 1: sections 1–4.

14. Homer, *Odyssey*, 1.1.

15. Zeus had designated *fertile Sicily to be the best of the fruitful earth*. [Pindar, *Nemean Odes*, in *Nemean Odes; Isthmian Odes; Fragments*, translated by William Race (Cambridge: Harvard University Press, 1997), 1.15.]

16. [Apollonius Rhodius, *Argonautica*, 4.964]; the triangular island of Thrinacia, or Sicily, is mentioned in Homer's *Odyssey* and Apollonius Rhodius' *Argonautica*. The early Greeks often referred to the constellation Triangulum as Deltaton, because of the similar isosceles-triangular shape of the letter Delta. [Allen, *Star Names*, p. 415]; Aratus described the constellation as *drawn with three sides, whereof two appear equal but the third is less*. [Aratus, *Phaenomena*, 234]; other observers referred to the constellation by various related titles, including Trinacia and Triqueta— terms that also referred to the triangular island of Sicily. [Callimachus, *Hymns and Epigrams*, in *Callimachus; Lycophron; Aratus*, translated by A. W. Mair (Cambridge: Harvard University Press, 1955), Hymn 3 to Artemis, 57, note *a*.]

17. Craters were typically used for diluting wine with water.

CHAPTER 3

1. *Homeric Hymns* describe the rugged wilds of ancient Arcadia as having *mountain peaks . . . rocky tracks... thick brush and towering crags*. [*Homeric Hymns; Homeric Apocrypha; Lives of Homer*, translated by Martin West (Cambridge: Harvard University Press, 2003), section 19.7.]

2. Eratosthenes credited Hesiod with this story. [Eratosthenes, *Catasterismi*, in *Star Myths of the Greeks and Romans*, compiled by Pseudo-Eratosthenes, translated by Theony Condos (Grand Rapids: Phanes Press, 1997), p. 197]; The Greek word for *bear* is *arktos*, from which the Arctic region gets its name. The Bear's unusually long tail has been variously explained. Some say that Zeus stretched the tail when he flung her into heaven. Because of the constellation's immense age, and its near-universal recognition as a bear (and usually a she-bear), some scholars suggest that Ursa Major dates to the Bering Strait migrations of the last Ice Age. Scholars have confirmed that many Native Americans knew the

constellation as a bear before the arrival of European influence. But the tail was interpreted differently, typically as three hunters, the middle one carrying a container—the dim star Alcor—on his back.

3. The name Bootes derives from *boetes*, which is Greek for *shouter*. [Allen, *Star Names*, 93.]

4. The Bear and the Little Bear are the modern constellations Ursa Major and Ursa Minor; but they are commonly known by the names of their asterisms: the Big Dipper and Little Dipper. The seven stars of each asterism can be seen within their respective constellations on the Northern Hemisphere Charts. Aratus called Ursa Minor the *Dog-tailed Bear*. [Aratus, *Phaenomena*, 182, 227.]

5. Aratus, *Phaenomena*, 276.

6. [Manilius, *Astronomica*, 4.521]; the five Hyades were the daughters of Atlas and Aethra. Zeus placed them in heaven in appreciation for their faithful service as nurses to the infant Dionysus. Zeus also sought in this way to comfort their undying grief over the death of their brother Hyas—a famed hunter killed by a wild boar. Both Homer and Hesiod mentioned the Hyades. Hesiod referred to them as *Nymphs similar to the Graces, Phaesyle and Coronis and well-garlanded Cleeia and lovely Phaeo and long-robed Eudora, whom the tribes of human beings on the earth call the Hyades*. [Hesiod, *Fragments*, in *The Shield; Catalogue of Women; Other Fragments*, translated by Glenn W. Most (Cambridge: Harvard University Press, 2007), section 227a.]

7. Apollonius Rhodius provided this tribute to Ariadne: *Once upon a time Ariadne, Minos' maiden daughter, rescued Theseus . . . from terrible trials through her kindness. Thereafter, she boarded his ship with him and left her country; and even the immortals themselves loved her, and in the midst of the sky her sign, a crown of stars they call Ariadne's, turns all night among the heavenly constellations*. [Apollonius Rhodius, *Argonautica*, 3.997]; Aratus called Corona Borealis *that Crown, which glorious Dionysus set to be memorial of the dead Ariadne*. [Aratus, *Phaenomena*, 71.]

CHAPTER 4

1. Homer, *Odyssey*, 11.309, 572.

2. Xenophon, *On Hunting*, in *Scripta Minora*, translated by E. C. Marchant (Cambridge: Harvard University Press, 1925), 1.18.

3. Xenophon, *On Hunting*, 4.1–2, 7–8; 5.1–5; 6.2–4, 11, 13.

4. Ibid, 4.4–5; 5.10, 16–17; 6.11, 13.

5. The parents, Atlas and Pleione, are now included in the asterism of the Pleiades, making a total of nine. But only six are easily visible. Of the sisters, Alcyone was the mother of Hyrieus (some say Arethusa) by Poseidon. Maia was the oldest and most beautiful, and probably the brightest star at one time (before that star's magnitude diminished). She was the mother of the god Hermes, by Zeus. Electra was the mother of Dardanus, founder of the royal house of

Troy, by Zeus. Merope was the wife of Sisyphus, King of Corinth. Taygeta was the mother of Lacedaemon, founder of Sparta, by Zeus. On either side of Taygeta, Celaeno and Sterope are the fainter and lesser known sisters. Athena's three temples on the Athenian Acropolis, dated 1530 BC, 1150 BC (the Hecatompedon), and 438 BC (the Parthenon), were all oriented toward the rising of the Pleiades, as were several other Greek temples. [Evans, *Ancient Astronomy*, p. 399.]

6. Aesop, *Fables*, Fable 5: The Fighting Cocks.

7. Aratus, *Phaenomena*, 402, 643.

8. Ancient Greek astronomers consistently viewed this constellation as the Claws of Scorpius. But later Roman astronomers began to refer to the constellation as a set of scales, or a balance. Libra first appeared as such in the Julian Calendar of 46 BC. The Roman astronomer Manilius wrote that when the sun is in Libra, *then the Balance, having matched daylight with the length of night, draws on the Scorpion.* [Manilius, *Astronomica*, 1.266]; in other words, the days and nights are equally balanced because, at the time of the autumnal equinox, the sun is over the equator. Eventually, the lethal claws of Scorpius became recognized as Virgo's scales of justice. Arabic astronomers, however, retained the Greek designation of the claws. They called two of the stars: Zubenelgenubi—the *southern claw* and Zubeneschamali—the *northern claw*. These remain their modern names.

9. Aratus, *Phaenomena*, 323.

10. The constellations Orion and Taurus represented to Sumerians a contest between Gilgamesh and the Bull of Heaven, as described in the *Epic of Gilgamesh*.

11. Homer, *Iliad*, 18.487.

12. According to Hyginus, Lepus is the eternal warning against introducing a prolific species into a new environment. In ancient times, hares were introduced onto the island of Leros with devastating results. With no natural predators to stop them, they soon ate all the crops. [Hyginus, *Poetic Astronomy*, in *Star Myths of the Greeks and Romans*, translated by Theony Condos (Grand Rapids: Phanes Press, 1997), p. 130.]

13. The Greeks referred to Canis Major as a spotted dog. Perhaps this is because some of his bright stars appear like spots on the rump. Some have suggested that the spots refer to his position in the speckled Milky Way, or are due to Sirius' strong variability as it rises in the atmosphere. [Allen, *Star Names*, pp. 119, 127]; Homer said that Sirius *shines brightest of all others when he has bathed in the stream of Ocean.* [Homer, *Iliad*, 5.1.]

14. The Greek term *Procyon* literally means *Herald of the Dog*, and at one time was applied to both the constellation and its brightest star. The ancient term has no association with the modern genus name of *Procyon*, which naturalists applied, in 1780, to the raccoon. This New World animal was first described by Christopher Columbus two thousand years after the era of classical Greece.

15. [Hesiod, *Theogony*, 338]; Phoebus means *bright* and is a common forename for Apollo.

16. Ovid, *Metaphorphoses*, translated by Frank Justus Miller (Cambridge: Harvard University Press, 1984), 2.319.

17. [Aratus, *Phaenomena*, 359; Apollonius Rhodius, *Argonautica*, 4.603]; the star called Achernar now marks the source of the river Eridanus. But because it appears far to the celestial south, it remained unknown to the ancient Greeks.

18. Homer, *Iliad*, 6.182.

19. Pindar, *Isthmian Odes*, 7.44–47.

20. Homer, *Iliad*, 6.200.

CHAPTER 5

1. Pegasus is depicted in the heavens as he emerges newborn from the sea foam, in the same manner that Taurus and Aries appear half-submerged in the sea.

2. Euripides, *Fragments*, in *Fragments: Aegeus to Meleager*, translated by Christopher Collard and Martin Cropp (Cambridge: Harvard University Press, 2008), Andromeda section 124.

3. Ovid, *Metamorphoses*, 4.625.

4. Ibid, 4.673.

5. Poseidon's wife Amphitrite was a Nereid, as was Achilles' mother, Thetis.

6. Euripides, *Fragments*, Andromeda section 145.

7. The Arabs later called the star Al-gol, meaning *the ghoul*.

8. Herodotus, *Histories*, 7: section 61.

9. Aratus, *Phaenomena*, 653.

10. Hesiod, *Theogony*, 278.

11. A winged horse, or *wind horse*, also appears in the religious symbolism of central and eastern Asia where it represents the soul. It is a common feature on Tibetan Buddhist prayer flags.

12. Celeris, during his lifetime, was a remarkably fast horse. Hermes, the messenger god, was so impressed with the speed of Celeris that he gave him to Castor, the greatest of mortal horsemen.

13. Heracles is also known by the Roman name Hercules.

14. Eratosthenes and Hyginus linked the constellation Hydra to the story of Corvus and Crater, but this is not attested in Archaic literature. Moreover, Aratus made no connection between Hydra, Corvus, and Crater, except that Corvus pecks at the serpent. Hesiod related the story of Hydra and Heracles at a much earlier date: *They say that Typhon, terrible, outrageous, lawless* sired *the evil-minded Hydra of Lerna* that Hera sent against Heracles. [Hesiod, *Theogony*, 306.]

15. The Hesperides were the three daughters of Atlas and Hesperis. They lived in the westernmost region of the known world, on the shores of the Strait of

Gibraltar. The names of the mother and daughters mean *western* or *evening*, denoting the evening setting of the sun in the west. Hesperis' father was Hesperos, the evening star. Our modern word *vesper* comes from this name.

16. Serpents are often depicted as trusted guardians, typically keeping watch over springs, homes, and temples. The Erechtheum on the Acropolis had its own guardian serpent that was tended by temple maidens. The constellation Draco was thought to never set in the west because he remained ever vigilant. One of Draco's stars, Thuban, marked the celestial north pole in 2750 BC. As such, all the stars in the sky rotated around it. This made Thuban and the constellation Draco extremely important in the ancient Near East. Thuban was visible from dusk to dawn from a main passage of the Great Pyramid of Cheops at Giza, at 30 degrees N latitude, and could be similarly viewed from the entrances of other sacred structures. [Allen, *Star Names*, pp. 206–207.]

CHAPTER 6

1. Apollodorus provided a concise account of the voyage, based upon Apollonius Rhodius. [Apollodorus, *The Library*, translated by James George Frazer (Cambridge: Harvard University Press, 1939), I.9, sections 16–28.]

2. [Apollonius Rhodius, *Argonautica*, 1.106, 4.1466]; the helmsman served as the primary navigator. He was the onboard expert in noting celestial signs both day and night. He interpreted the winds, ripples, currents, and flocks of migrating birds. His mind was a storehouse of information on coastal landmarks, dangerous obstacles, and the local aggregate of the seafloor. He also steered the ship and adjusted the sail by means of brail lines. As such, he remained indispensable. [Samuel Mark, *Homeric Seafaring* (College Station: Texas A&M Press, 2005), p. 148.]

3. Apollonius Rhodius, *Argonautica*, 4.1192.

4. Ibid, 2.160.

5. Ibid, 3.189, note 10.

6. Ibid, 2.1145.

7. Ibid, 4.123.

8. Standard shipbuilding tools of ancient Greece included an axe and an adze for cutting and shaping timbers and mortises; an auger for drilling holes for pegs; a mallet for joining the timbers by means of pegs and mortises; and a chalk line to shape the timbers straight and true. [For descriptions and illustrations of tools used for constructing boats and dwellings, see Anthony Rich, *The Illustrated Companion to the Latin Dictionary and Greek Lexicon* (London: Longman, Brown, Green, and Longmans, 1849); W. M. Flinders-Petrie, *Tools and Weapons* (London: British School of Archaeology in Egypt, 1917); Henry Mercer, *Ancient Carpenters' Tools* (Doylestown, PA: Horizon Press, 1975)]; Homer described the tools that Calypso provided to Odysseus so he could build a raft: *She gave him a*

big axe, well fitted to his hands, an axe of bronze, sharpened on both sides; and in it was a beautiful handle of olive wood, securely fastened; and thereafter she gave him a polished adze.... Twenty trees in all did he fell, and trimmed them with the axe; then he cunningly smoothed them all and trued them to the line. Meanwhile Calypso, the beautiful goddess, brought him augers; and he bored all the pieces and fitted them to one another, and with pegs and mortising he hammered it together.... And he set in place the decks, bolting them to the close-set ribs, as he continued the work; and he finished the raft with long gunwales. In it he set a mast and a yardarm, fitted to it, and furthermore made him a steering oar.... Meanwhile Calypso, the beautiful goddess, brought him cloth to make him a sail, and he fashioned that too with skill. And he made fast in the raft braces and halyards and sheets, and then with levers worked it down into the bright sea. [Homer, *Odyssey*, 5.233.]

9. Apollonius Rhodius, *Argonautica*, 1.1182.

10. Ibid, 1.562.

11. Pindar, *Pythian Odes*, in *Olympian Odes; Pythian Odes*, translated by William Race (Cambridge: Harvard University Press, 1997), 4.202–203.

12. Apollonius Rhodius, *Argonautica*, 1.1272.

13. [Apollonius Rhodius, *Argonautica*, 1.1084]; a halcyon was probably a kingfisher. Alcyone, the Pleiad, got her name from this bird.

14. Apollonius Rhodius, *Argonautica*, 1.1118, note 118.

15. The Black Sea was once known as the Pontus. Propontis means *Before Pontus* and is now called the Sea of Marmara.

16. Apollonius Rhodius, *Argonautica*, 3.957.

17. Ibid, 3.744.

18. Ibid, 3.1348.

19. Ibid, 3.1372.

20. Pindar said the serpent *exceeded in breadth and length a ship of fifty-oars.* [Pindar, *Pythian Odes*, 4.244–245.]

21. The Aries motif appeared in ancient Greek art, most notably in images of Hermes Kriophoros (the ram-bearer), and remained popular into late antiquity. The Ram looking back over his flank, as in the constellation image, became a common pose that continued to appear in Christian depictions of the Good Shepherd in sculptures of the first centuries AD, and in stucco in the Catacomb of Priscilla, Rome, and elsewhere.

22. Homer revered the ship of the Argonauts as *that Argo famed of all.* [Homer, *Odyssey*, 12.70]; because of its size in the heavens, Argo Navis was later divided into four constellations: Carina, Vela, Pyxis, and Puppis. It also contained one star now part of the modern constellation Columba.

23. Pindar, *Pythian Odes*, 4.176–177.

24. Apollonius Rhodius, *Argonautica*, 1.26.

25. Ibid, 1.496.

26. Ibid, 4.904.

27. Ibid, 4.1193.

28. Homer, *Odyssey*, 8.478.

29. Pindar said, Chiron *raised Jason in his rocky dwelling and then Asklepios, whom he taught the gentle-handed province of medicines.* [Pindar, *Nemean Odes*, 3.53–55.]

30. For one of many modern depictions of this *Rod of Asclepius*, see the flag of the United Nations World Health Organization.

31. Herodotus, *Histories*, 8: section 41.

32. Apollonius Rhodius, *Argonautica*, 4.592.

33. Ibid, 4.652.

34. The star that marks Polydeuces' head is known in modern times by his Latin name: Pollux.

35. Ancient sources refer to these appearances of St. Elmo's Fire. One of the *Homeric Hymns* declared that *when winter tempests race over the implacable sea, and the men from their ships invoke the Sons of great Zeus in prayer, with [sacrifice of] white lambs, going onto the stern deck, and the strong wind and sea swell overwhelm the ship: suddenly they [Castor and Polydeuces] appear, speeding through the air on tawny wings, and at once they make the fierce squalls cease, and lay the waves amid the flats of a clear sea—fair portents, and release from travail; the sailors rejoice at the sight, and their misery and stress are ended.* [*Homeric Hymns*, section 33]; Alcaeus wrote praises to the Twins who *rescue men from chilling death, leaping on the peaks of their well-benched ships, brilliant from afar as you run up the fore-stays, bringing light to the black ship in the night of trouble.* [Alcaeus, *Fragments*, in *Greek Lyric: Sappho; Alcaeus*, translated by David Campbell (Cambridge: Harvard University Press, 1982), Fragment 34]; Pliny also described this appearance of St. Elmo's Fire as the favorable light of Castor and Polydeuces: *On a voyage, stars alight on the yards and other parts of the ship, with a sound resembling a voice, hopping from perch to perch in the manner of birds.... If there are two of them, they denote safety and portend a successful voyage;... for this reason they are called Castor and Pollux, and people pray to them as gods for aid at sea.* [Pliny, *Natural History*, translated by H. Rackham (Cambridge: Harvard University Press, 1949), 2.37]; In like manner, the *Bible* recorded the high regard in which ancient seafarers held the Twins. On one occasion, the Apostle Paul was taken under military escort from Crete to Rome. Along the way, the crew encountered fourteen stormy nights at sea and was shipwrecked on the island of Malta. Paul and the Roman guards recuperated for three months before securing another ship to Rome. This time they boarded a ship that was sure to be safer because it had the sign of Castor and Pollux on the figurehead of the bow. [*Book of Acts*, 28:11]; according to James Evans, *the Greek Alexandria, and Ostia, the harbor of Rome,* were considered to be *under the tutelage of the Twins, who were often represented on either side of the bows of vessels owned in those ports.* [Evans, *Ancient Astronomy*, p. 226.]

PART TWO

1. Johann Bayer's constellations of 1603 adorn the ceiling of Grand Central Terminal in New York City.

CHAPTER 9

1. Homer, *Odyssey*, 24.143.

2. Aratus, *Phaenomena*, 740.

3. The Greeks marked the seasons by noting the positions of stars, asterisms, or constellations on the eastern and western horizons at dawn and dusk. At early dawn, while the stars still shine, the heliacal rising of a star denotes its first appearance in the east, just ahead of the rising sun. Every morning thereafter, the same star appears higher in the sky at dawn and farther ahead of the sun. Months later, the same star finally sets on the western horizon at dawn. This is called the apparent cosmical setting. The Greeks noted similar positions at dusk. The first appearance of a star on the eastern horizon at dusk is called the apparent achronycal rising; and the appearance of a star on the western horizon at dusk, as it follows the setting sun, is called the heliacal setting. After its heliacal setting, a star disappears from the night sky for several weeks and reappears in the morning sky at its heliacal rising. In summary, at dawn, the heliacal rising of a star occurs just ahead of the rising sun, while the apparent cosmical setting of another star is occurring opposite the sun. At dusk, the apparent achronycal rising of a star occurs opposite the setting sun, while the heliacal setting of another star is occurring just behind the setting sun. For example, in the time of Hesiod, the Pleiades made a heliacal rising in the east at dawn on May 16. After crossing the night sky for several months, the Pleiades made an apparent cosmical setting on November 3. Several months later, its heliacal setting occurred at dusk on April 5. At that point, the Pleiades disappeared from the sky for about forty nights until making a heliacal rising again on May 16. Twenty-seven centuries later, the precession of the Earth (the slow wobble around its axis) has caused these stars to appear several days later than in ancient times. Risings and settings are also affected by differences in latitude and several other factors.

4. Aratus, *Phaenomena*, 562.

5. All dates are approximations applied to our modern calendar. They are based on calculations and computer simulations for 700 BC, at the latitude of Athens. See Appendix 3: Annual Celestial Events, noted by Hesiod with additions by Eudoxus.

6. Hesiod, *Works and Days*, 414, 427.

7. [Hesiod, *Works and Days*, 423]; for illustrations and descriptions of ancient

tools, see Rich, *Illustrated Companion to the Latin Dictionary and Greek Lexicon*; and Flinders-Petrie, *Tools and Weapons*.

8. [Hesiod, *Works and Days*, 564]; when providing day counts, Hesiod rounded estimates to multiples of ten (e.g. *sixty wintry days*, and *forty nights and days*). [A. W. Mair, *Hesiod: The Poems and Fragments Done into English Prose* (Oxford: Clarendon Press, 1908), p. 144; Harald Reiche, "Fail-Safe Stellar Dating: Forgotten Phases," *Transactions of the American Philological Association* 119 (1989): 45–46.]

9. The pruning hook was sometimes used. It consisted of a sickle-shape iron blade and wooden handle—a smaller version of the reaping sickle used for the grain harvest. [Mair, *Hesiod*, p. 153.]

10. Homer, *Odyssey*, 24.246, 340.

11. Xenophon, *Oeconomicus*, in *Memorabilia and Oeconomicus*, translated by E. C. Marchant (Cambridge: Harvard University Press, 2013), 19.3–5, 7–8, 11.

12. Theophrastus, *Concerning Weather Signs*, in *Enquiry into Plants*, translated by Arthur Hort (Cambridge: Harvard University Press, 1949), 1.10, 14, 15; 3.38–40, 46, 47.

13. Xenophon, *Oeconomicus*, 17.12.

14. Hesiod, *Works and Days*, 569.

15. Aelian, *On the Characteristics of Animals*, translated by A. F. Scholfield (Cambridge: Harvard University Press, 1958), 1.52.

16. Aristotle, in typical manner, offered a scientific explanation for this migratory bird behavior: *Animals have an innate perception of change in respect of hot and cold. . . . Some find protection . . . in their habitual locations, while others [like cranes and fishes] migrate.* Aristotle refers here to the common crane—*Grus grus*. [Aristotle, *History of Animals*, translated by D. M. Balme (Cambridge: Harvard University Press, 1991), 709.]

17. *When the Atlas-born Pleiades rise, start the harvest—the plowing, when they set. They are concealed for forty nights and days, but when the year has revolved they appear once more, when the iron is being sharpened.* [Hesiod, *Works and Days*, 383, 571; Xenophon, *Oeconomicus*, 18.1–2.]

18. Hesiod, *Works and Days*, 597.

19. Xenophon, *Oeconomicus*, 18.1, 4–5.

20. In the land of Canaan, the threshing floor of Araunah, the Jebusite, was situated on a hilltop for this same purpose. When David, the neighboring Hebrew King, sought a lofty site to serve as a Temple Mount, he purchased Araunah's threshing floor. For the following three thousand years, Jews, Christians, and Muslims have cherished the sacred site. The Temple Mount, that was once a threshing floor, is now home to the mosque called Dome of the Rock, in Jerusalem. [*2 Samuel* 24:18–25, *1 Chronicles* 21:18–26.]

21. Xenophon, *Oeconomicus*, 18.7–8; Hesiod, *Works and Days*, 606.

22. Hesiod, *Works and Days*, 414; Apollonius Rhodius, *Argonautica*, 2.524.

23. Hesiod, *Works and Days*, 393, 582, 588.

24. Xenophon, *Oeconomicus*, 19.18–19.

25. Homer, *Odyssey*, 24.340.

26. Xenophon, *Oeconomicus*, 19.19.

27. Hesiod, *Works and Days*, 609.

28. Ibid, 383, 571, 615.

29. Ibid, 458, 479.

30. Theophrastus, *Concerning Weather Signs*, 4.55; Aratus, *Phaenomena*, 1064.

31. Aristotle, *History of Animals*, 597a.21.

32. Theophrastus, *Concerning Weather Signs*, 3.38; Aratus, *Phaenomena*, 1075; Hesiod, *Works and Days*, 448; Aristophanes, *Birds*, in *Birds; Lysistrata; Women at the Thesmophoria*, translated by Jeffrey Henderson (Cambridge: Harvard University Press, 2000), 709.

33. Hesiod, *Works and Days*, 427, 469.

34. Aristophanes, *Acharnians*, in *Acharnians; Knights*, translated by Jeffrey Henderson (Cambridge: Harvard University Press, 1998), 1025.

35. Hesiod, *Works and Days*, 427.

36. Xenophon, *Oeconomicus*, 17.2–4.

37. Ibid, 17.1–6.

38. Ibid, 17.7–8, 11.

39. Hesiod, *Works and Days*, 470; Xenophon, *Oeconomicus*, 5.6.

40. [*Leviticus* 26:5; *Deuteronomy* 11:14]; the farming cycle is further described by Mair in *Hesiod: The Poems and Fragments*.

41. Hesiod, *Works and Days*, 463.

42. Hesiod, *Works and Days*, 462; Xenophon, *Oeconomicus*, 16.11–15.

43. Hesiod, *Works and Days*, 551, 576.

44. The ancient Greeks celebrated agriculture as a gift from the gods. In particular, they praised Demeter—the goddess of farming and harvest—as the divine benefactress who first gave man the golden seeds of grain and taught him how to plant and harvest a crop. The stories of old told how she poured the priceless gift of grain into the hands of a pious man named Triptolemus. At the time, Triptolemus served as a priest at the shrine of Eleusis, near Athens, where he devoted his life to the ritual worship of Demeter. After receiving the grain, he followed her will in a more fruitful way. First, he planted, harvested, threshed, and winnowed a crop at Eleusis. Then, with bags of grain seed, he left the shrine behind and traveled the breadth of Greece to bestow the blessings of agriculture. In this manner, he offered the means of subsistence and survival to his countrymen and assured a steady supply of food for future generations. Through the centuries, Greeks continued to favor farming as a noble profession and a praiseworthy occupation. Xenophon voiced a common belief—repeated through the ages—that farmers, being bound to the land, have the greatest stake in their country, and the strongest desire to

defend it against invasion. Farming, he said, *seems to turn out citizens who are the bravest and most loyal to the community.* Xenophon also considered farming the most feasible means of assuring a family's survival. He claimed that it ranked as one of the healthiest and most rewarding ways to make a living: *The best occupation and the best branch of knowledge is farming, from which people obtain what is necessary to them. For this occupation seemed to be the easiest to learn and the most pleasant to practice; to afford the body the greatest measure of strength and beauty; and to afford the mind the greatest amount of spare time.* Any person, declared Xenophon, can learn to farm through common sense and simple observation of their surroundings. For example, if a person sees an uncultivated meadow producing happy, healthy native plants, then he knows, straight away, that the soil is fertile. A person may also note the types of plants that local farmers successfully grow. In this way, he will know at a glance what the soil can bear. *The land,* he said, *makes no deceptive displays but reveals frankly and truthfully what she can and cannot do.* Xenophon pointed out that, in addition to making simple observations, a person can seek the advice of agrarian neighbors. Almost every farmer proves ready and willing to offer his knowledge and local lore to all who will listen, so to acquire a basic knowledge of farming is a simple matter. Beyond that, a person must only be willing to work. The difference between a good farmer and a bad farmer, said Xenophon, is not so much due to a difference in knowledge as it is due to a difference in diligence. According to nature's design, the lazy person goes hungry, and the hard worker comes home to find food on the table. In the words of Hesiod: *Famine is ever the companion of a man who does not work.* Even a farmer of minimal financial means can succeed, said Xenophon, if he is willing to work. One way to begin, or to add to one's holdings, is to buy cheap land in a poor location, with poor soil, then work hard to improve it. If the land is low and boggy, it can be drained. If the soil is weak or alkaline, it can be strengthened with richer soil, and with fertilization from manure or vegetative mulch. Xenophon concluded that in farming, it is possible for any determined person to learn the profession and to succeed at the art of subsistence and survival. [*Homeric Hymns,* "To Demeter," 2.474; Apollodorus, *The Library,* 1.5.2; Pausanias, *Description of Greece,* "Attica," 14.2–4; 38.6–7; Hesiod, *Works and Days,* 302; Xenophon, *Oeconomicus,* 6.8–10; 15.10–12; 16.3, 5; 18.10; 20.2, 6, 11–13, 22–23.]

45. Aristophanes, *Birds,* 713.
46. Theocritus, *Idylls,* in *Theocritus; Moschus; Bion,* translated by Neil Hopkinson (Cambridge: Harvard University Press, 2015), 13.25.
47. Sophocles, *Oedipus Tyrannus,* in *Ajax; Electra; Oedipus Tyrannus,* translated by Hugh Lloyd-Jones (Cambridge: Harvard University Press, 1994), 1137.
48. Aelian, *On the Characteristics of Animals,* 7.8.
49. Apollonius Rhodius, *Argonautica,* 4.1629.

50. Sappho, *Fragments*, in *Greek Lyric: Sappho; Alcaeus*, translated by David Campbell (Cambridge: Harvard University Press, 1982), Fragment 104.

51. Hesiod, *Theogony*, 22; Albert Schachter, *Cults of Boiotia* (London: University of London, Institute of Classical Studies, 1986).

52. Homer, *Iliad*, 8.555.

53. [Theocritus, *Idylls*, 13.25]; likewise, the Greek fleet of a thousand ships that sailed to Troy—a city strategically located near the entrance of the Hellespont—gathered first and wintered at Aulis (modern Avlida) in Boeotia. Only after winter's end did they dare to cross the Aegean Sea. [Homer, *Iliad*, 2.303; Hesiod, *Works and Days*, 651.]

54. Hesiod, *Works and Days*, 663.

55. Ibid, 618.

56. Apollonius Rhodius, *Argonautica*, 1.1203.

57. Aratus, *Phaenomena*, 167.

58. Hesiod, *Works and Days*, 623.

CHAPTER 10

1. Morton's excellent study is reflected in this chapter. [Jamie Morton, *The Role of the Physical Environment in Ancient Greek Seafaring* (Leiden: Brill, 2001).]

2. Hesiod, *Works and Days*, 678.

3. Leonidas of Tarentum, *Epigrams*, in *The Greek Anthology*, translated by W. R. Paton (Cambridge: Harvard University Press, 1948), 10.1.

4. Marcus Argentarius, *Epigrams*, in *The Greek Anthology*, translated by W. R. Paton (Cambridge: Harvard University Press, 1948), 10.4.

5. Strabo, *Geography*, in *The Geography of Strabo*, translated by Horace L. Jones (Cambridge: Harvard University Press, 1949), 10.4.5.

6. Homer, *Odyssey*, 14.252.

7. See Appendix 5, Aegean Sea Map.

8. Aristotle, *History of Animals*, 596b.24.

9. Aristophanes, *Birds*, 709.

10. Theocritus, *Epigrams*, in *Theocritus; Moschus; Bion*, translated by Neil Hopkinson (Cambridge: Harvard University Press, 2015), 25.3.

11. Xenophon described the restless nature of merchant sailors as they fervently searched for the most prized commodity—grain: *On receiving reports that it is abundant anywhere, merchants will voyage in quest of it: they will cross the Aegean, the Euxine [Black], the Sicilian sea, and when they have got as much as possible, they carry it over the sea and actually stow it in the very ship in which they themselves sail. And when they need money they don't unload the grain just anywhere, but they carry it to the place where they hear that grain is most valued.* [Xenophon, *Oeconomicus*, 20.27.]

12. The name came from the belief that the halcyon seabird had the power to calm the waters. Theocritus claimed that halcyons, being the *birds dearest to the sea-*

green Nereids and to fishermen, would often *smooth the sea*. [Theocritus, *Idylls*, 7.57 and note 15.]

13. Aratus, *Phaenomena*, 758.

14. Homer, *Odyssey*, 12.310.

15. Theophrastus, *Concerning Weather Signs*, 1.10, 2.27, 3.38.

16. Aratus, *Phaenomena*, 408.

17. Theophrastus, *Concerning Weather Signs*, 1.22; 2.32, 34; 3.43, 45.

18. Ibid, 2.31, 35, 37.

19. Aratus, *Phaenomena*, 1010; Theophrastus, *Concerning Weather Signs*, 4.52; Aelian, *On the Characteristics of Animals*, 3.13.

20. Theophrastus, *Concerning Weather Signs*, 1.20; 3.38, 40; Aelian, *On the Characteristics of Animals*, 3.14, 7.7.

21. Greeks on land and sea had long observed the four primary winds according to their most common properties: the colder north wind, hotter south wind, wetter west wind, and drier east wind. Their sensory observations of temperature and humidity worked in ways that seem amazing today.

22. Homer, *Odyssey*, 24.76.

23. Strabo, *Geography*, 3.2.5.

24. A stadion was a unit of measure in the ancient Mediterranean world. It was based on the pous (pl. podes)—our equivalent of the foot. A pous consisted of sixteen finger widths. With an average of about three-fourths inch per finger width, the pous measured some twelve inches. One hundred podes equaled a plethron (pl. plethra), and six plethra equaled a stadion. The stadion (stadium in Latin) was the length of the most popular foot race, and the standard length of the edifice (stadium) in which it was run. Stadion distances varied by locale. In general, a stadion measured about six hundred feet, or 183 meters.

25. Strabo, *Geography*, 6.3.10, 10.4.5.

26. Anaximander (c. 610–c. 546 BC) is credited with bringing the gnomon to the Greeks. This simple vertical stick was used in his time to cast the sun's shadow in order to determine noon. Also, the length of the noon shadow indicated the solstices. [Robert Hannah, *Time in Antiquity* (London: Routledge, 2009), p. 69]; Pytheas of Massalia (present-day Marseilles), a Greek navigator of the fourth century BC, put the gnomon to good use when he brazenly sailed from the Mediterranean Sea into the rough waters of the Atlantic Ocean, then ventured northward to the British Isles and beyond.

27. Homer, *Odyssey*, 5.269.

28. Thales of Miletus (c. 624–546 BC), promoted the use of Ursa Minor and convinced Greek mariners to follow the more accurate Phoenician method. [Callimachus, *Iambi*, in *Aetia; Iambi; Lyric Poems*, translated by C. A. Trypanis (Cambridge: Harvard University Press, 1958), 1.52.]

29. Eratosthenes, *Catasterismi*, p. 201.

30. Aratus, *Phaenomena*, 37.

31. Phoenician sailors enjoyed a widespread reputation as excellent mariners. Accordingly, Xenophon expressed amazement at the impeccable outfitting of their merchant ships: *I have never seen tackle so excellently and accurately arranged.* Each piece rested in its own compact storage receptacle, ready for immediate use. This included the full array of rigging, as well as a worthy assortment of weapons for fighting pirates. It also consisted of utensils for cooking and eating, and all the cargo carried for trade: *Each kind of thing was so neatly stowed away that there was no confusion, no work for a searcher, nothing out of place, no troublesome untying to cause delay when anything was needed for immediate use.* Moreover, the captain's first mate routinely checked the quantity and arrangement of equipment and supplies. In their small, wooden world, order was essential: *People aboard a merchant vessel, even if it's a little one, find room for things and keep order, though tossed violently to and fro, and find what they want to get, though terror stricken.* [Xenophon, *Oeconomicus*, 8.11–17.]

32. In addition to his talents as a mathematician, Eratosthenes was an astronomer, geographer, historian, music theorist, poet, and athlete. He coined the term *geography* and founded the discipline. He created a map of the known world, and invented latitude and longitude. He proved the Earth's spherical shape, calculated its circumference, and determined its approximate axis tilt. As an astronomer, he invented the armillary sphere and compiled a star catalog—the *Catasterismi*. As a historian, he recorded a chronology from the fall of Troy to his own time. As a humanitarian, he condemned Aristotle's division of humans into Greeks and barbarians. He also produced a work on ethics. He served as the third head librarian at Alexandria, succeeding Apollonius Rhodius. His close friend was the famed mathematician and inventor, Archimedes.

33. Aratus, *Phaenomena*, 727.

34. Homer, *Odyssey*, 12.310, 14.482.

APPENDIX 3

1. Hesiod used five easily identifiable celestial objects, also mentioned by Homer, to provide nine calendar markers of practical value to farmers, shepherds, and sailors. He also used the summer and winter solstices to mark general seasonal characteristics. He was aware of the equinoxes (Hesiod, *Fragments*, 226), but found little calendar use for them as they proved difficult to pinpoint. Eudoxus' contributions to Hesiod's calendar are preserved by Aratus in the *Phaenomena*. His supplements include the four constellation markers of the solstices and equinoxes; additional information added to five of Hesiod's calendar events; and three new markers. The approximate calendar dates shown here are based on calculations for the year 700 BC at the latitude of Athens. The dates are provided courtesy of Anthony B. Kaye, PhD, professor of physics and astronomy at Texas Tech University. The following is a summary of the complexities

and methodologies involved in producing these calculations: Gregorian calendar dates for the vernal equinox, autumnal equinox, summer solstice, and winter solstice were all computed using the algorithm given in Jean Meeus, *Astronomical Algorithms*, 2nd ed. (Richmond, Virginia: Willmann-Bell, Inc., 1998). This returned a Julian date as a solution, which was converted to a Gregorian calendar date using the "Julian Date Converter" application made available by the United States Naval Observatory. The determination of the other four quantities (heliacal rising, heliacal setting, apparent achronycal rising, and apparent cosmical setting) proved more complicated. To compute these dates, one must first compute the positions of each of the stars in the sky. As each star has its own three-dimensional coordinates (given by its right ascension, declination, and distance), it also has three-dimensional motion in the sky (given by its proper motion in right ascension and declination and its radial velocity). Thus, one may go back in time to determine the positions of the stars given their current positions (or rather, their positions on a given day) and their individual motions. However, the problem is not that simple. Due to the c. 2700 years considered here, the complicated motion of the Earth has to be taken into account. A good discussion of the various effects can be found in James G. Williams, "Contributions to the Earth's Obliquity Rate, Precession, and Nutation," *Astronomical Journal* 108 (1994): 711–724. The three main effects on the Earth's motion, in addition to its simple revolution around the sun, are precession, nutation, and obliquity. Precession is a result of the gravity of the sun and the moon acting on the oblateness of the Earth. As a result, the Earth's spin axis describes a circle of roughly 23.4 degrees and takes c. 26,000 years to complete. The consequence of precession is that the north star—Polaris—is drifting. During the time period considered here, the north celestial pole was remarkably dark. The closest bright star would have been Kochab, in Ursa Minor. Nutation is an effect caused mainly by the precession of the planet on the moon's tilted, elliptical orbit. This means that there is another change in the direction of the north celestial pole, but with a smaller amplitude and a shorter period of 18.6 years. Obliquity, which is a measure of the tilt of the Earth's axis, ranges a total of plus or minus 1.3 degrees from its average value of 23.4 degrees. Further, since the night sky appears differently from each location on Earth, we must have reliable coordinates, including an altitude, for the observer. In addition to these "external" considerations, there are also more "local" variables that must be addressed. These include questions like, Where is the observer on the Earth? At what altitude? How dark is the night sky in that location and at that time? [See: R. H. Garstang, "Night-Sky Brightnesses at Observatories and Sites," *Publications of the Astronomical Society of the Pacific* 101 (1989): 306–329; and K. Krisciunas, "Further Measurements of Extinction and Sky Brightness on the Island of Hawaii," *Publications of the Astronomical Society of the Pacific* 102 (1990): 1052–1063.] Other questions include, What was the atmosphere like? Was it stable, or was it turbulent enough to cause starlight to shimmer drastically? What were the effects of atmospheric refraction and aberration? What was the weather like that day? (In addition to clouds, mist, and fog, local temperature and humidity also affected the specific time of the event.) How experienced was the observer? How good was his vision? Of course, these latter questions cannot be answered, so we must work with optimal viewing conditions. [See Bradley Schaefer, "Predicting Heliacal Risings and Settings," *Sky & Telescope* 70 (1985): 261–263; G. V. Rozenberg, "Light Scattering in the Earth's Atmosphere," *Soviet Physics Uspekhi* 3 (1960): 346–371; H. Richard Blackwell, "Contrast Thresholds of the Human Eye," *Journal of the Optical Society of America* 36 (1946): 624–643; and I. S. Bowen, "Limiting Visual Magnitude," *Publications of the Astronomical Society of the Pacific* 59 (1947): 253–256.] Further, although we have various definitions of different kinds of twilight today, those definitions did not exist in the first millennium BC. So, if we use the term "heliacal rising" to mean the rising of a star with or just before the sun, how close can the object be to the sun before the sun's light overwhelms it and makes it invisible? With all of these variables, we decided to use four different methods to calculate the four events of interest: 1) The technique of Lacroix—a loop-based method that marks when the star and the sun are at altitudes that allow first visibility of the star; 2) The method described by Karine Gadré, "Conception d'un modèle de visibilitè d'ètoile à l'oeil nu. Application à l'identification des dècans ègyptiens" (PhD dissertation, Universitè Paul Sabatier, 2008); 3) The software *Planetary, Lunar, and Stellar Visibility* (PLSV), version 2.0, which is based on a slightly more elaborate version of Ptolemy's method described in the *Almagest*; and, 4) The software *Stellarium* (v 0.16) with the "Observability Analysis" plugin (v 1.0.2).

2. Hesiod, *Works and Days*, 564–570.

3. [Aratus, *Phaenomena*, 745]; Aratus said *dread Arcturus* marks stormy seasons with both its heliacal and its apparent achronycal risings.

4. Aratus, *Phaenomena*, 516.

5. Hesiod, *Works and Days*, 384.

6. Ibid, 383, 572.

7. Aratus, *Phaenomena*, 266.

8. [Hesiod, *Works and Days*, 597]; Hesiod said winnowing begins *when Orion's strength first shows itself.* Orion's rising and setting is marked by the star Betelgeuse because it is the first bright star of the constellation to rise and the last to set.

9. Hesiod, *Works and Days*, 663.

10. Aratus, *Phaenomena*, 500.

11. [Hesiod, *Works and Days*, 587; Aratus, *Phaenomena*, 332]; Hesiod does not specify the time of Sirius' rising. Aratus is more specific.

12. Aratus called the Lion *the Sun's hottest summer path.* [Aratus, *Phaenomena*, 151];

Apollonius said the Etesian Winds last forty days. [Apollonius Rhodius, *Argonautica*, 2.524.]

13. [Aratus, *Phaenomena*, 138]; Aratus does not define the role of this star because the name requires no further explanation.

14. [Hesiod, *Works and Days*, 610]; Sirius is used as an approximate marker of the grape harvest at this time, as it approaches a southern zenith at dawn. Homer and Hesiod both note its prominent appearance during the harvest. [Homer, *Iliad*, 5.5, 22.26; Hesiod, *Works and Days*, 609.]

15. Aratus, *Phaenomena*, 745.

16. [Aratus, *Phaenomena*, 519, 546]; Aratus does not specifically mention the Claws as an equinoctial marker, but makes the constellation's role obvious in his designation of the twelve zodiacal constellations.

17. Hesiod, *Works and Days*, 383, 615, 619.

18. [Aratus, *Phaenomena*, 266]; Aratus agrees that this marks the plowing season.

19. Hesiod, *Works and Day*, 615.

20. [Hesiod, *Works and Days*, 615]; the Pleiades, Hyades, and Orion offer three distinct warnings to complete the plowing before winter. Otherwise, the crop will fail. [Hesiod, *Works and Days*, 479.]

21. [Aratus, *Phaenomena*, 313]; Aratus calls Eagle the *Storm-bird*, that *tosses in the storm . . . in its rising from the sea when the night is waning.*

22. Hesiod, *Works and Days*, 564.

23. Aratus, *Phaenomena*, 284, 501.

WORKS CITED

Ancient Greek constellation mythology and lore spans nine hundred years of literature, from Homer to Claudius Ptolemy. Many ancient sources, including the works of Eudoxus and Hipparchus, are tragically lost. But some are preserved in paraphrase or fragmentary form in later works. For example, we can trace Aratus' *Phenomena* and Eratosthenes' *Catasterismi* back to Eudoxus. And we know that Eudoxus, in the fourth century BC, described almost all the forty-eight constellations later recorded by Ptolemy in the second century AD.

Aratus based his astronomical poem on the work of Eudoxus, but offered little more than fleeting information for each constellation. Eratosthenes followed Eudoxus more closely. He compiled the earliest comprehensive collection of constellation stories that exists in basic form today. While Eratosthenes' original work—*Catasterismi*—vanished in antiquity, a later compilation of his work remains. It is called by the same name and written by an unknown author often labeled Pseudo-Eratosthenes.

Homer, engraved on a bronze coin of Smyrna, second century BC. Marshall Collection, Loma Paloma, Texas. Photo by author.

Hyginus was the only other ancient author to preserve the constellation stories in substantial form. His *Poetic Astronomy* reflects Eratosthenes in form and content. Thus, the works of Hyginus and Pseudo-Eratosthenes represent a continuity of thought dating back to Eratosthenes and Eudoxus. Together, the four span the period from Classical Greece to the Roman Empire.

The most important ancient sources on constellation mythology are:

Homer (c. 750 BC)
Hesiod (c. 700 BC)
Eudoxus (c. 410–355 BC), who influenced Aratus and Eratosthenes
Aratus (c. 315–240 BC)
Apollonius Rhodius (c. 310–246 BC)
Eratosthenes (c. 276–195 BC), who influenced Hyginus and Pseudo-Eratosthenes
Hipparchus (c. 190–120 BC), who influenced Claudius Ptolemy and the Farnese Atlas sculptor

Apollodorus (c. 180–120 BC)
Hyginus (c. 64 BC–17 AD)
Pseudo-Eratosthenes (c. 50 BC)
Claudius Ptolemy (c. 90–168 AD)
The Farnese Atlas sculptor (c. 150 AD)

Claudius Ptolemy, engraved on a silver plate, sixth century AD. Courtesy of the John Paul Getty Museum, Getty Villa, Malibu, California. Photo by author.

PRIMARY SOURCES

Aelian. *On the Characteristics of Animals.* Translated by A. F. Scholfield. Cambridge: Harvard University Press, 1958.

Aesop. *Fables.* In *Babrius and Phaedrus.* Compiled by Babrius. Translated by Ben Edwin Perry. Cambridge: Harvard University Press, 1965.

Alcaeus. *Fragments.* In *Greek Lyric: Sappho; Alcaeus.* Translated by David Campbell. Cambridge: Harvard University Press, 1982.

Anaximander. *Fragments.* In *Early Greek Philosophy.* Quoted by Theophrastus and Simplicius. Translated by John Burnet. London: Adam and Charles Black, 1930.

Apollodorus. *The Library.* Translated by James George Frazer. Cambridge: Harvard University Press, 1939.

Apollonius Rhodius. *The Argonautica.* Translated by William H. Race. Cambridge: Harvard University Press, 2008.

Aratus. *Phaenomena.* In *Callimachus; Lycophron; Aratus.* Translated by G. R. Mair. Cambridge: Harvard University Press, 1955.

Aristophanes. *Acharnians.* In *Acharnians; Knights.* Translated by Jeffrey Henderson. Cambridge: Harvard University Press, 1998.

———. *Birds.* In *Birds; Lysistrata; Women at the Thesmophoria.* Translated by Jeffrey Henderson. Cambridge: Harvard University Press, 2000.

Aristotle. *History of Animals.* Translated by D. M. Balme. Cambridge: Harvard University Press, 1991.

———. *Metaphysics.* Translated by Hugh Tredennick. Cambridge: Harvard University Press, 1947.

Callimachus. *Hymns and Epigrams*. In *Callimachus; Lycophron; Aratus*. Translated by A. W. Mair. Cambridge: Harvard University Press, 1955.

———. *Iambi*. In *Aetia; Iambi; Lyric Poems*. Translated by C. A. Trypanis. Cambridge: Harvard University Press, 1958.

Epic of Gilgamesh. Translated by N. K. Sandars. New York: Penguin Books, 1972.

Eratosthenes. *Catasterismi*. In *Star Myths of the Greeks and Romans*. Compiled by Pseudo-Eratosthenes. Translated by Theony Condos. Grand Rapids: Phanes Press, 1997.

Euripides. *Fragments*. In *Fragments: Aegeus to Meleager*. Translated by Christopher Collard and Martin Cropp. Cambridge: Harvard University Press, 2008.

Heraclitus. *Fragments*. In *Ancilla to the Pre-Socratic Philosophers*. Quoted by Hippolytus. Translated by Kathleen Freeman. Cambridge: Harvard University Press, 1962.

Herodotus. *The Histories*. Translated by Robin Waterfield. Oxford: Oxford University Press, 1998.

Hesiod. *Fragments*. In *The Shield; Catalogue of Women; Other Fragments*. Translated by Glenn W. Most. Cambridge: Harvard University Press, 2007.

———. *Testimonia*. In *Theogony; Works and Days; Testimonia*. Translated by Glenn W. Most. Cambridge: Harvard University Press, 2006.

———. *Theogony*. In *Theogony; Works and Days; Testimonia*. Translated by Glenn W. Most. Cambridge: Harvard University Press, 2006.

———. *Works and Days*. In *Theogony; Works and Days; Testimonia*. Translated by Glenn W. Most. Cambridge: Harvard University Press, 2006.

Homer. *Iliad*. Translated by A. T. Murray and William F. Wyatt. Cambridge: Harvard University Press, 1999.

———. *Odyssey*. Translated by A. T. Murray and George E. Dimock. Cambridge: Harvard University Press, 1995.

Homeric Hymns; Homeric Apocrypha; Lives of Homer. Translated by Martin West. Cambridge: Harvard University Press, 2003.

Hyginus. *Poetic Astronomy*. In *Star Myths of the Greeks and Romans*. Translated by Theony Condos. Grand Rapids: Phanes Press, 1997.

Leonidas of Tarentum. *Epigrams*. In *The Greek Anthology*. Translated by W. R. Paton. Cambridge: Harvard University Press, 1948.

Manilius. *Astronomica*. Translated by G. P. Goold. Cambridge: Harvard University Press, 1977.

Marcus Argentarius. *Epigrams*. In *The Greek Anthology*. Translated by W. R. Paton. Cambridge: Harvard University Press, 1948.

Ovid. *Metamorphoses*. Translated by Frank J. Miller. Cambridge: Harvard University Press, 1984.

Pausanias. *Description of Greece*. Translated by W. H. S. Jones. Cambridge: Harvard University Press, 1955.

Pherecydes. *Fragments*. In *Pherekydes of Syros*. Quoted by Proclus. Translated by Hermann Schibli. Oxford: Clarendon Press, 1990.

Pindar. *Isthmian Odes*. In *Nemean Odes; Isthmian Odes; Fragments*. Translated by William Race. Cambridge: Harvard University Press, 1997.

———. *Nemean Odes*. In *Nemean Odes; Isthmian Odes; Fragments*. Translated by William Race. Cambridge: Harvard University Press, 1997.

———. *Pythian Odes*. In *Olympian Odes; Pythian Odes*. Translated by William Race. Cambridge: Harvard University Press, 1997.

Plato. *Phaedo*. In *The Dialogues of Plato*. Translated by B. Jowett. New York: Random House, 1937.

———. *Protagoras*. In *The Dialogues of Plato*. Translated by B. Jowett. New York: Random House, 1937.

———. *The Republic*. Translated by Paul Shorey. Cambridge: Harvard University Press, 1970.

———. *Timaeus*. In *The Dialogues of Plato*. Translated by B. Jowett. New York: Random House, 1937.

Pliny. *Natural History*. Translated by H. Rackham. Cambridge: Harvard University Press, 1949.

Ptolemy. *Almagest*. In *Syntaxis Mathematica*. Translated by J. L. Heiberg. Leipzig: B. G. Teubneri, 1898.

———. *Almagest*. In *Ptolemy's Almagest*. Translated by G. J. Toomer. Princeton: Princeton University Press, 1998.

Sappho. *Fragments*. In *Greek Lyric: Sappho; Alcaeus*. Translated by David Campbell. Cambridge: Harvard University Press, 1982.

Sophocles. *Oedipus Tyrannus*. In *Ajax; Electra; Oedipus Tyrannus*. Translated by Hugh Lloyd-Jones. Cambridge: Harvard University Press, 1994.

Strabo. *Geography*. In *The Geography of Strabo*. Translated by Horace L. Jones. Cambridge: Harvard University Press, 1949.

Theocritus. *Epigrams*. In *Theocritus; Moschus; Bion*. Translated by Neil Hopkinson. Cambridge: Harvard University Press, 2015.

———. *Idylls*. In *Theocritus; Moschus; Bion*. Translated by Neil Hopkinson. Cambridge: Harvard University Press, 2015.

Theophrastus. *Concerning Weather Signs*. In *Enquiry into Plants*. Translated by Arthur Hort. Cambridge: Harvard University Press, 1949.

Xenophanes. *Fragments*. In *Ancilla to the Pre-Socratic Philosophers*. Quoted by Clement and Sextus Empiricus. Translated by Kathleen Freeman. Cambridge: Harvard University Press, 1962.

Xenophon. *Oeconomicus*. In *Memorabilia and Oeconomicus*. Translated by E. C. Marchant. Cambridge: Harvard University Press, 2013.

———. *On Hunting*. In *Scripta Minora*. Translated by E. C. Marchant. Cambridge: Harvard University Press, 1925.

SECONDARY SOURCES

Allen, Richard. *Star Names: Their Lore and Meaning*. New York: Dover Publications, 1963.

Diels, Hermann. *Die Fragmente der Vorsokratiker*. Berlin: Weidmann, 1964.

Evans, James. *The History and Practice of Ancient Astronomy*. Oxford: Oxford University Press, 1998.

Flinders-Petrie, W. M. *Tools and Weapons*. London: British School of Archaeology in Egypt, 1917.

Freeman, Kathleen. *Ancilla to the Pre-Socratic Philosophers*. Cambridge: Harvard University Press, 1962.

Gibbon, William. "Asiatic Parallels in North American Star Lore: Milky Way, Pleiades, Orion." *Journal of American Folklore*. 85 no. 337 (1972): 236–247.

———. "Asiatic Parallels in North American Star Lore: Ursa Major." *Journal of American Folklore*. 77 no. 305 (1964): 236–250.

Hannah, Robert. *Time in Antiquity*. London: Routledge, 2009.

Mair, A. W. *Hesiod: The Poems and Fragments*. Oxford: Clarendon Press, 1908.

Makemson, Maud. "Astronomy in Primitive Religion." *Journal of Bible and Religion*. 22 no. 3 (1954): 163–171.

Mark, Samuel. *Homeric Seafaring*. College Station: Texas A&M Press, 2005.

Mercer, Henry. *Ancient Carpenters' Tools*. Doylestown, PA: Horizon Press, 1975.

Morton, Jamie. *The Role of the Physical Environment in Ancient Greek Seafaring*. Leiden: Brill, 2001.

Rappenglueck, Michael. "The Pleiades in the 'Salle des Taureaux,' Grotte de Lascaux," www.infis.org/research.

Reiche, Harald. "Fail-Safe Stellar Dating: Forgotten Phases," *Transactions of the American Philological Association* 119 (1989): 37–53.

Rich, Anthony. *The Illustrated Companion to the Latin Dictionary and Greek Lexicon*. London: Longman, Brown, Green, and Longmans, 1849.

Schachter, Albert. *Cults of Boiotia*. London: University of London, Institute of Classical Studies, 1986.

INDEX

Note: *Italicized* pages refer to photos or illustrations; **bolded** pages refer to constellation charts.